W9-CEA-669

Project Earth Science:
Physical Oceanography

From *The Sargasso Sea* by John and Mildred Teal. Copyright © 1975 by John and Mildred Teal. Reprinted by permission of authors.

"Sound of Water" from WHAT IS THAT SOUND! by Mary O'Neill. Copyright © 1966 by Mary O'Neill. Reprinted by permission of Marian Reiner.

From "The Rime of the Ancient Mariner" by Samuel Taylor Coleridge. Material is in the public domain.

"A Wet and A Flowing Sea" by unknown. Material is in the public domain.

"Little Things" by Julia Carney. Material is in the public domain.

"Neither Out Far Nor In Deep" by Robert Frost. From THE POETRY OF ROBERT FROST edited by Edward Connery Lathem. Copyright © 1936 by Robert Frost. Copyright © by Lesley Frost Ballantine. Copyright © 1969 by Henry Holt and Co., Inc. Reprinted by permission of Henry Holt and Co., Inc.

From "Rockall" by Epes Sargent. Material is in the public domain..

From "Apostrophe to the Ocean" by Lord Byron. Material is in the public domain.

From "The Odyssey" by Homer. Material is in the public domain.

"Sailing" by unknown. Material is in the public domain.

"There's a Hole in the Bottom of the Sea" by unknown. Material is in the public domain.

"Until I Saw the Sea" from I FEEL THE SAME WAY by Lilian Moore. Copyright © 1967 by Lilian Moore. Reprinted by permission of Marian Reiner for the author.

From "The Forsaken Merman" by Matthew Arnold. Material is in the public domain.

"The Tide" by Henry Wadsworth Longfellow. Material is in the public domain.

From "Water" by Kathleen Raine. Material is in the public domain.

From "The Pleasure Boat" by Richard Henry Dana. Material is in the public domain.

"Strandloper Stones" by Dorian Haarhoff from *Acquifers and Dust*. Reprinted by permission of Justified Press, a division of William Waterman Publications, P.O. Box 5091, Rivonia, 2128, South Africa.

"On the Beach at Night Alone" by Walt Whitman. Material is in the public domain.

Project Earth Science: Physical Oceanography

by

Brent A. Ford and P. Sean Smith

Now featuring *sci*LINKS® a new way of connecting text and the Internet. Up-to-the-minute online content, classroom ideas, and other materials are just a click away. Go to page 210 to learn more about this new educational resource.

NATIONAL SCIENCE TEACHERS ASSOCIATION

A Project of Horizon Research, Inc.

Material for middle school teachers in Earth science

This project was funded by BP America, Inc. and published by the National Science Teachers Association

Project Earth Science: Physical Oceanography

Second Edition

Table of Contents

Acknowledgments 7

Introduction 9

    Overview of Project Earth Science 9

    About *Project Earth Science:*
    Physical Oceanography 10

    Getting Ready for Classroom Instruction 10

    Key Concepts 11

    Project Earth Science: Physcial Oceanography and
    the *National Science Education Standards* 13

Activities

Activity 1   A Pile of Water 15

Activity 2   A Sticky Molecule 25

Activity 3   Over and Under—Why Water's Weird 35

Activity 4   How Water Holds Heat 41

Activity 5   Water—The Universal Solvent 49

Activity 6   Won't You BB My Hydrometer? 57

Activity 7   Ocean Layers 67

Activity 8   The Myth of Davy Jones's Locker 75

Activity 9   Estuaries—Where the Rivers Meet the Sea 85

NATIONAL SCIENCE TEACHERS ASSOCIATION

Current Events in the Ocean 91   **Activity 10**

Body Waves 101   **Demonstration 11**

Waves and Wind in a Box 107   **Activity 12**

Tanks A Lot—Activities for a Wave Tank 115   **Demonstration 13**

Plotting Tidal Curves 125   **Activity 14**

Tides Mobile 133   **Activity 15**

The Bulge on the Other Side of Earth 141   **Demonstration 16**

Oily Spills 147   **Activity 17**

Forever Trash 155   **Activity 18**

Readings

Water: The Sum of Its Parts 163   **Reading 1**

The Ocean 171   **Reading 2**

The Tides: A Balance of Forces 177   **Reading 3**

Waves 185   **Reading 4**

The Ocean: A Global View 189   **Reading 5**

Appendices

Appendix A: Master Materials List 195

Appendix B: Resource Guide 198

Appendix C: Constructing a Wave Tank 203

Appendix D: *Standards* Organizational Matrix .. 208

Appendix E: *sci*LINKS® 210

Shirley Watt Ireton, Director
Beth Daniels, Managing Editor
Jessica Green, Assistant Editor
Anne Early, Editorial Assistant

Art and Design
Kim Alberto, Director
NSTA Web
Tim Weber, Webmaster
Outreach
Michael Byrnes, Editor-at-Large
Periodicals Publishing
Shelley Carey, Director
Printing and Production
Catherine Lorrain-Hale, Director
Publications Operations
Erin Miller, Manager
sci*LINKS*
Tyson Brown, Manager

National Science Teachers Association
Gerald F. Wheeler, Executive Director
David Beacom, Publisher

NSTA Press, NSTA Journals, and the NSTA Web site deliver high-quality resources for science educators.

About the cover images
Front Cover:
• A shaded relief image of the Pacific Ocean basin. The darker the blue, the deeper the water. Courtesy National Oceanic and Atmospheric Administration.
• Columbia Glacier ice off the coast of Alaska. Photo by Kurt Byers, courtesy Alaska Sea Grant College Program.

Back Cover:
• This nautical chart shows where the Atlantic Ocean meets the Virginia shoreline. The numbers indicate water depth in feet. From the *Guide for Cruising Maryland Waters*, courtesy Maryland Department of Natural Resources.

Acknowledgements

Numerous people have contributed to the development of *Project Earth Science: Physical Oceanography*. The volume began as a collection of activities and readings from Project Earth Science, a teacher enhancement project funded by the National Science Foundation. Project Earth Science was designed to provide in-service education for middle school Earth science teachers in North Carolina. Nine two-person leadership teams received extensive training in conducting in-service workshops on selected topics in astronomy, geology, meteorology, and oceanography. They in turn conducted in-service education programs for teachers throughout the state of North Carolina. Principal investigators for this project were Iris R. Weiss, president of Horizon Research, Inc.; Diana Montgomery, research associate at Horizon Research, Inc.; Paul B. Hounshell, professor of education, University of North Carolina-Chapel Hill; and Paul Fullagar, professor of geology, University of North Carolina-Chapel Hill.

The activities and readings have undergone many revisions as a result of comments and suggestions provided primarily by the Project Earth Science leaders, and also by workshop participants, project consultants, and project staff. Project leaders include: Kevin Barnard, Winston–Salem/Forsyth County Schools; Kathy Bobay, Charlotte-Mecklenburg Schools; Pam Bookout, Guilford County Schools; Betty Dean, Guilford County Schools; Lynanne (Missy) Gabriel, Charlotte-Mecklenburg Schools; Flo Gullickson, Guilford County Schools; Michele Heath, Chapel Hill/Carrboro Schools; Cameron Holbrook, Winston-Salem/Forsyth County Schools; Linda Hollingsworth, Randolph County Schools; Geoff Holt, Wake County Schools; Kim Kelly, Chapel Hill/Carrboro City Schools; Laura Kolb, Wake County Schools; Karen Kozel, Durham County Schools; Kim Taylor, Durham County Schools; Dana White, Wake County Schools; Tammy Williams, Guilford County Schools; and Lowell Zeigler, Wake County Schools.

Special thanks to the following: Kim Taylor for her contributions to "A Pile of Water," "A Sticky Molecule," "Over and Under—Why Water's Weird," "How Water Holds Heat," "Water—The Universal Solvent," and "Forever Trash;" Jack E. Gartrell, Jr., a consultant to Horizon Research Inc., for his contributions to "Won't You BB My Hydrometer?,"

"Estuaries—Where the Rivers Meet the Sea." and "Current Events in the Ocean;" Linda Ford, Vice President of Novostar Designs, Inc., for significant contributions in the form of writing and editing on the readings and many of the activities; Dirk Frankenberg, Professor of Marine Sciences at University of North Carolina-Chapel Hill, for contributing valuable suggestions on all activities; and Shirley Brown, a teacher with the Columbus City Schools in Ohio and a participant in the Program for Leadership in Earth Systems Education (PLESE) at Ohio State University, for contributing many of the ideas in "Suggestions for Interdisciplinary Study" and finding poetry to accompany the activities.

Thanks to Kim Taylor, who contributed significantly in the preparation of the bibliography and to Kevin Barnard, Kathy Bobay, Missy Gabriel, Linda Hollingsworth, and Tammy Williams, who assisted in annotating the citations.

During the publication process, excellent reviewers provided comments and suggestions to NSTA: Richard Benz, Ron Morse, Susan Leach Snyder, Carolyn Staudt, Keith Sverdrup, and Ken Wardwell.

Project Earth Science: Physical Oceanography was produced by NSTA Special Publications. Shirley Watt Ireton, managing editor; Joe Cain, associate editor; Gregg Sekscienski, associate editor; Michelle Eugeni, program assistant; Glen Fullmer; editorial assistant. Gregg Sekscienski was project editor for *Project Earth Science: Physical Oceanography*. Illustrations were created by the authors and Linda Ford, and by Max-Karl Winkler. The book was designed by Marty Ittner of AURAS Design and printed by Automated Graphic Systems, Inc.

Special thanks go to BP America, Inc., for providing the funds to make this book possible.

Introduction

Project Earth Science: Physical Oceanography is the third in a four-volume series of Earth science books. The other volumes are *Project Earth Science: Astronomy*, *Project Earth Science: Meteorology*, and *Project Earth Science: Geology*. Each book contains a collection of hands-on activities developed for the middle/junior high school level and a series of background readings pertaining to the topic area.

Overview of Project Earth Science

Project Earth Science was a teacher enhancement program initially funded by the National Science Foundation. Originally conceived as a program in leadership development, this project was designed to prepare middle school teachers to lead workshops on topics in Earth science. These workshops were designed to help teachers convey key Earth science concepts and content through the use of hands-on activities. With the help of content-area experts, concept outlines were developed for specific topics and activities were designed to illustrate the concepts. Some activities were drawn from existing sources, and many others were developed by Project Earth Science leaders and staff.

Over the course of the two-year project, the activities were field-tested in workshops and classrooms, and extensively revised. After each revision, the activities were reviewed by content experts for accuracy. Finally, all activities were organized in a standard format. During the publication process, the book underwent another extensive review and revision.

The grand theme of the Project Earth Science materials is the uniqueness of Earth among all the planets in the Solar System. Concepts and activities were chosen that elaborate on this theme. The *Physical Oceanography* volume focuses on the uniqueness of water, the effects of its special characteristics on Earth's ocean, and the forces that cause currents, waves, and tides. The effect that humans have on the marine environment is also addressed.

About *Project Earth Science: Physical Oceanography*

This book is divided into three sections: activities, readings, and appendices. The activities in this volume are organized under three broad concepts. First, students investigate the unique properties of water and how these properties shape the ocean and the global environment. Second, students perform activities investigating the complex systems that lead to the development of currents, waves, and tides. This section focuses on the interactions of wind, water, gravity, and inertia. In the third section, students study the impact that humans have on the ocean and the marine environment, particularly effects of pollutants.

A set of readings follows the activities. Some readings are intended to enhance teacher preparation—or serve as additional resources for students interested in further study—by elaborating on concepts presented in the activities. Other readings introduce supplemental material so that teachers can connect science to broader social issues.

An annotated bibliography is included as Appendix B and is intended to serve as a supplemental materials guide. Entries are divided into various categories: activities and curriculum projects; books and booklets; audiovisual materials; instructional aids; information and references; Sea Grant programs, and Internet resources.

Getting Ready for Classroom Instruction

The activities in this volume are designed to be hands-on and can be performed using materials that are either readily available in the classroom or inexpensive to purchase. Each activity has a student section—ready for duplication—and a teacher's guide. The student section contains background information that introduces the concept behind each activity. Following this is the procedure for the activity and a set of questions to guide the students as they draw conclusions.

The teacher versions of the activities contain a more detailed version of the background information and a summary of the important points that students should understand after completing the activity. You'll find the approximate time to allow for each activity in the "Time Management" section. The "Preparation" section describes the set-up for the activity and gives sources of materials for some activities. "Suggestions for Further Study" offers ideas for challenging students to extend their study of the topic. Use these ideas within the class period allotted for the activity if time and enthusiasm allow. "Suggestions for Interdisciplinary Reading and Study" includes ideas for relating the concepts in the activity to other science topics and to other disciplines, such as language arts and social studies. The final section provides answers to the questions in the student section.

In addition, each activity begins with a poem or quote related to the ocean. The sea is an eternal source of fascination and inspiration to scientific and artistic intellects alike. You may want to copy these poems to accompany the activities—maybe to remind students of the interwoven nature of arts and sciences or to use elsewhere in the curriculum.

Project Earth Science began as a workshop method to provide assistance to teachers on Earth science content and instructional techniques. To use this volume for teacher workshops, we have included a master materials list (see Appendix A).

Key Concepts

The activities are organized around three key concepts: the investigation of water and its properties; the forces that affect water's movement on Earth; and the human impact on Earth's ocean. The presentation of concepts and participation in activities should be an integrated process. To aid in this coordination, a conceptual outline for *Project Earth Science: Physical Oceanography* is presented below:

1. Although water is a common substance, many of its familiar characteristics make it unique among molecules. Its special properties lead to the characteristics of Earth's ocean which make the planet uniquely life-bearing among its neighbors in the solar system. An awareness of the chemical structure and physical characteristics of water underlies an understanding of its importance on Earth.

Activities:
A Pile of Water
A Sticky Molecule
Over and Under—Why Water's Weird
How Water Holds Heat
Water—The Universal Solvent

2. Movement of water within the ocean occurs through the development of currents, waves, and tides. Deep ocean currents are caused by variations in ocean water density; surface currents result mainly from wind. Waves represent energy in motion and result primarily from the wind as well. Tides are produced by the interaction of forces involving Earth, the Sun, and the Moon.

Activities:
Won't You BB My Hydrometer?
Ocean Layers
The Myth of Davy Jones's Locker
Estuaries: Where the Rivers Meet the Sea
Current Events in the Ocean
Body Waves
Waves and Wind in a Box
Tanks A Lot—Activities for a Wave Tank
Plotting Tidal Curves
Tides Mobile
The Bulge on the Other Side of Earth

3. Human activities have an impact on Earth's ocean. The effects are long-lasting and sometimes irreversible.

Activities:
Oily Spills
Forever Trash

Project Earth Science: Physical Oceanography and the National Science Education Standards

Effective science teaching within the middle-level age cluster integrates the two broadest groupings of scientific activity identified by the *National Science Education Standards*: (1) developing skills and abilities necessary to perform scientific inquiry, and (2) developing an understanding of the implications and applications of scientific inquiry. Within the context of these two broad groupings, the *Standards* identify specific categories of classroom activity that will encourage and enable students to integrate skills and abilities with understanding.

To facilitate this integration, an organizational matrix for *Project Earth Science: Physical Oceanography* appears on pages 208–209. The categories listed along the X-axis of the matrix, also listed below, correspond to the categories of performing and understanding scientific activity identified as appropriate by the *Standards*.

Subject Matter and Content. Specifies the topic covered by an activity.

Scientific Inquiry. Identifies the "processes of science" (ie. scientific reasoning, critical thinking, conducting investigations, formulating hypotheses) employed by an activity.

Unifying Concepts and Processes. Links an activity's specific subject topic with "the big picture" of scientific ideas (ie. how data collection techniques inform interpretation and analysis).

Technology. Establishes a connection between the natural and designed worlds.

Personal/Social Perspectives. Locates the specific meteorology topic covered by an activity within a framework that relates directly to students' lives.

By integrating the presentation of specific science subject matter with the encouragement of students to organize and locate that subject matter within an accessible framework, *Project Earth Science: Physical Oceanography* hopes to address the *Standards'* call for making science—in this case oceanography—something students do, not something that is done to students. The organizational matrix provides a tool to assist teachers in realizing this goal.

from *The Sargasso Sea*

The planet Earth is misnamed, if you
consider it from the proportion of land
to water. By a fair naming process
it would be called Water.

John and Mildred Teal

A Pile of Water

Background

We observe and use water every day. It makes life on Earth possible. Water covers nearly three-fourths of Earth's surface and affects almost all living and non-living things. Because it is so abundant, it may not seem unusual, but water is unique when compared to other substances in the universe. As a matter of fact, its properties are quite different from those of other substances even here on Earth. For instance, it is the only substance on Earth that occurs naturally in all three states—solid (an iceberg, for example), liquid (ocean water), and gas (steam or vapor). See Figure 1.

Every substance (water, air, dirt, etc.) is made up of atoms. Atoms are arranged in a specific way, forming a **molecule**. The make-up of a water molecule—or any molecule—is called **molecular structure**. A substance's molecular structure is responsible for its properties and governs how it interacts with other things on Earth. This activity introduces and explores one specific property of liquid water.

Objective

The purpose of this activity is to investigate a specific property of water: its ability to "stick" to itself.

Vocabulary

Molecule: The most basic unit of a substance; it has a specific arrangement of atoms.

Molecular structure: The arrangement of atoms in a specific molecule.

Figure 1

In areas like this arctic region water can often be found in all three states.

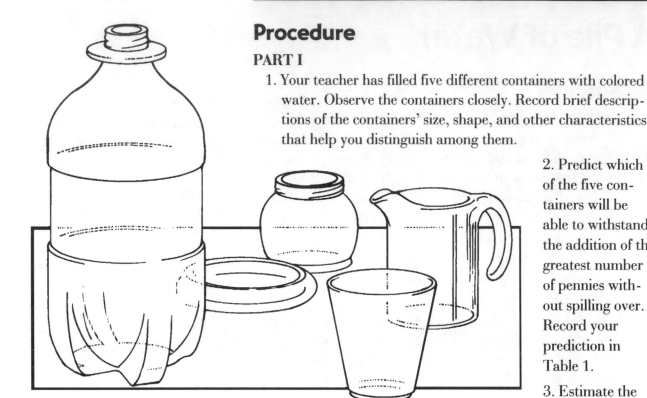

Procedure

PART I

1. Your teacher has filled five different containers with colored water. Observe the containers closely. Record brief descriptions of the containers' size, shape, and other characteristics that help you distinguish among them.

2. Predict which of the five containers will be able to withstand the addition of the greatest number of pennies without spilling over. Record your prediction in Table 1.

3. Estimate the number of pennies that would have to be added to each container to make it spill over. Record your estimates in Table 1. You will have a chance to revise these estimates in step 7 of Part II.

PART II

Before you check your predictions, explore the behavior of water with a single coin.

1. Place a penny on a piece of paper towel.

2. Estimate the number of water drops you can pile on the penny before the water runs over its edge. Record your estimate in Table 2.

3. Test your estimate by placing water on the penny drop by drop. Working with the other members of your group, develop a technique that allows you to put the most drops on your penny. You may want to put the drops on at different areas of the penny or from different heights. Count each drop until the water spills over. Record your results in Table 2.

4. Make a sketch of the water on the surface of the penny just before the water spilled over.

5. Based on what you observed with the penny, make a hypothesis comparing the number of drops that could be piled on a nickel, dime, or quarter.

Materials

For each group (class will be divided into five groups):

◊ 1 of 5 containers varying in shape and size of opening

◊ several rolls of pennies

For each student:

◊ penny

◊ eyedropper

◊ small beaker or clear plastic cup (may be shared)

◊ paper towel

◊ other assorted coins (quarters, nickels, etc.)

DATA TABLE 1: INITIAL PREDICTIONS

Container predicted to hold the most pennies before overflowing:

Container number	Description	Number of pennies predicted to cause overflow
1		
2		
3		
4		
5		

6. Repeat steps 2, 3, and 4 with each coin to test your hypothesis.

7. Look back at the predictions you made in Part I. Based on the experimenting you have just completed, would you like to change any of your predictions? Record any revised estimates in Table 3. Why did you choose to change or not to change your predictions?

8. Working with your group, test your hypothesis about one of the five containers from Part I and record your results. Use the following procedure to add pennies to a container until it overflows.

a. Hold the penny so its edge will enter the water first (not flat).

b. Hold the penny no more than 5 cm above the surface of the water over the center of the container opening.

c. Release the penny and let it drop into the water.

DATA TABLE 2: PREDICTED AND ACTUAL RESULTS

Item	Number of drops	
	Prediction	Actual
Penny		
Nickel		
Dime		
Quarter		

DATA TABLE 3: REVISED PREDICTIONS AND ACTUAL RESULTS

Container predicted to hold the most pennies before overflowing:

Container number	Number of pennies predicted to cause overflow	Number of pennies required to cause overflow
1		
2		
3		
4		
5		

9. Compare your hypotheses for the other containers with the results obtained by the other groups. Also compare the techniques you devised that allowed you to put the most drops on a coin.

Questions/Conclusions

1. Describe the way water "sits" on the penny.

2. Why do some pennies hold more water droplets than others?

3. Why do you think water piles up on the penny, rather than spilling over the edges immediately?

4. Suggest reasons why the five containers hold a different number of pennies before spilling over.

5. In Step 7, did you change your predictions from Part I before testing your original prediction? Describe how this is consistent with the way scientists test their ideas.

A Pile of Water

What is Happening?

The properties of water play an integral role in the development and maintenance of Earth's environment and its ability to sustain life. Although water is a unique substance in the universe, it is so common on Earth that many students may expect other substances to have properties similar to water.

Water exhibits characteristics that are unusual compared to other substances. For instance, solid water floats in liquid water (most solids of a substance are more dense than the liquid and therefore sink), large amounts of energy must be added to water to achieve relatively small changes in temperature (heat capacity), and water molecules tend to "stick" to each other (cohesion) and to other molecules (adhesion). Later activities will explore the first two properties while this activity introduces students to liquid water's ability to "stick" to itself.

Water molecules "stick," or are attracted to one another, because water is a polar molecule with an uneven distribution of electrical charge. Each molecule has a positive end or "pole" and a negative pole. The positive end of one molecule and the negative end of another molecule attract each other. This attraction, called hydrogen bonding, is strong enough to hold water molecules together. The force of hydrogen bonds causes water to fall in drops and to dome up on flat surfaces or containers full of water.

When placed on coins, the molecules of water form flexible piles which stay together because of hydrogen bonding. This phenomenon—of water "piling up"— is called surface tension. Liquid water has an extremely high surface tension because of its molecular structure and the hydrogen bonding between molecules.

Important Points for Students to Understand

◊ Water is unique among substances.
◊ The high surface tension of water, which results in water "piling up" on a flat surface, is just one of the unusual properties of water. Water's characteristics are important in determining how water interacts with other substances.

Materials

For teacher (for preparation ahead of activity):
◊ food coloring

◊ water

For each group (class will be divided into five groups):
◊ 1 of 5 containers varying in shape and size of opening

◊ several rolls of pennies

For each student:
◊ penny

◊ eyedropper

◊ small beaker or clear plastic cup (may be shared)

◊ paper towel

◊ other assorted coins (quarters, nickels, etc.)

◊ The characteristics of water make life possible on Earth. One of them, surface tension, was investigated in this activity. It allows water striders to "walk" on the surface of water. It also aids in capillary action, which allows groundwater to move through soil, wells to function, and plants to transport water from their roots upward.

Time Management

The observation of the five different containers (Part I) as well as the experimentation with the coins and containers (Part II) may be completed in one class period.

Preparation

Choose five containers, each differing in volume and mouth shape and size. The containers may be of like or different materials (plastic, glass, etc.). Possible containers might include an apple juice jar, a wide-mouthed plastic cup, a petri dish, a two-liter plastic drink bottle, and a canning jar. It is important that the containers chosen for the activity have a variety of opening sizes, from small to large. *The success of the activity depends on using a wide variety of containers.* All containers should be transparent.

Fill each container with water and add a different color to each. Be sure all containers are filled to the point where the water is exactly level with the opening of the container. The water should not form a depression (meniscus) or be domed up in the center of the opening. You might want to do this part of the activity ahead of time.

Give students an opportunity to record their results on the board and make comparisons with other groups in the class. Have students compute a class average for the results from Table 2, then construct two bar charts: one tallying student's predictions and the other tallying the actual number of drops that fit on each coin. What conclusions emerge from looking at the data in this way? (Did most students predict too many drops or too few?)

Suggestions for Further Study

Challenge students to float a paper clip on a petri dish full of water. With the paper clip floating, add a drop or two of liquid detergent or soap. Watch what happens and ask students to try to re-float the paper clip. Soap reduces the surface tension of the water, making the paper clip sink.

Compare the surface tension of tap water and salt water. Although the addition of impurities—like salt—decreases the cohesion between water molecules, it also increases the density of water (density will be explored in Activity 3). For example, the presence of large quantities of salt allows objects that would sink in fresh water to float on the surface of water in the Dead Sea and the Great Salt Lake. This may be confusing to students. The ability to float results from a change in density rather than an increase in surface tension.

Challenge students to repeat Part II with liquids other than water. Their predictions before testing will expose any misconceptions that all liquids behave as water does. Rubbing alcohol, white vinegar, and vegetable oil make good alternatives.

Explore capillary action—the mechanism by which groundwater moves and trees and plants transport water from their roots throughout the plant—and "water striders" (animals that take advantage of water's surface tension to live on the surface of the water).

Answers to Questions for Students

1. The water forms a dome on the surface of the penny. (NOTE: The hydrogen bonds hold the water molecules together, allowing many more drops than would be expected to pile on the penny.)

2. The pennies hold different numbers of drops because they are not all exactly the same. Some pennies have worn edges, others are dirty or dented. There may also be a difference between the heads and the tail sides of the pennies. (NOTE: There also may be differences in the size of the drops. Encourage students to develop techniques that "standardize" their drop size.)

3. Answers will vary since students have not been introduced to hydrogen bonding. Students may create explanations, though, that are fairly accurate. Encourage students to formulate hypotheses to explain why the water piled up. Address these hypotheses through discussion. The answer: The hydrogen bonding between water molecules hold the molecules together. When the number of water molecules gets too large for the cohesive forces to hold together, the water spills over the side.

4. Again, answers here will vary. Ask students to formulate and discuss their hypotheses. The openings or mouths of the

containers differ in surface area and shape. Containers with large round openings will hold a greater number of pennies than those with small round openings. There may also be differences in the adhesive forces between water and the material from which the containers are made.

5. Answers will vary. (NOTE: Most students greatly underestimate the number of pennies it takes to make the water spill over the sides of the containers. Therefore, after observing the water dome up on the coins, most students will increase their estimates.) Scientists often form their initial hypothesis based on intuition or intitial observations. Hypotheses are revised as additional data are collected through experimentation or further observations.

Sound of Water

The sound of water is:
Rain,
Lap,
Fold,
Slap,
Gurgle,
Splash,
Churn,
Crash,
Murmur,
Pour,
Ripple,
Roar,
Plunge,
Drip,
Spout,
Skip,
Sprinkle,
Flow,
Ice,
Snow.

Mary O'Neill

A Sticky Molecule

Background

Water is one of the simplest substances on Earth, and yet we must have it to live. A person can survive only about three days without water. Water is so common that we often ignore some of its characteristics. Pure water is clear, with no odor or taste. These are three obvious properties of water, others are less obvious. For example, compared to other liquids, it takes a lot of heat to make water hot and even more to make it boil.

Knowing the molecular structure of water—or any other susbtance—helps in understanding many of its properties. In every water molecule, two hydrogen atoms are joined to one oxygen atom by forces called **chemical bonds**. The chemical bond between hydrogen and oxygen happens when the two atoms *share* two electrons between them—one electron from hydrogen and one from oxygen. This is shown in Figure 1.

In this activity, you will learn how hydrogen and oxygen join and investigate some characteristics of the bond between them. This will help you understand the properties of water— many of which you will investigate in later activities.

Objective

The purpose of this activity is to construct a model of the water molecule, in order to explore the concepts of polarity and hydrogen bonding.

Materials

For each student:
◊ paper molecule pattern
◊ scissors
◊ glue
◊ crayons or markers (red and blue)
◊ blank paper

Vocabulary

Chemical bonds: The forces between atoms that hold a molecule together.

Procedure

1. Locate the Water Molecule Pattern Sheet. Color both hydrogen atoms and nuclei blue and the oxygen atom and nucleus red, leaving the electrons uncolored. Cut out all the pieces (atoms, nuclei, and electrons) of the water molecule.

2. Remember that in every water molecule, two hydrogen atoms are bonded to one oxygen atom. Before gluing the hydrogen atoms to the oxygen atom, try different arrange-ments of the atoms and use a blank piece of paper to sketch out at least three different ways that the two hydrogen atoms and one oxygen could be joined in a single molecule.

3. Earlier this century, scientists discov-ered the actual shape of a water molecule. The hydrogen atoms are attached to the oxygen atom in a

Figure 1

The bond between hydrogen and oxygen occurs when they share two electrons—one from hydrogen and one from oxygen

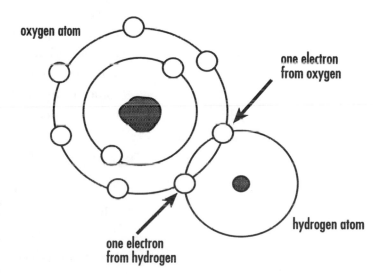

oxygen atom

one electron from oxygen

one electron from hydrogen

hydrogen atom

way that makes the molecule look a lot like the head of a mouse if you viewed a mouse head-on. The oxygen atom is the mouse's head and the two hydrogen atoms are its ears.

4. Based on this description, glue the hydrogen and oxygen atoms together. The glue represents the bonding between the hydrogen atoms and the oxygen atom.

5. Glue a nucleus in the center of each atom so that the nucleus covers the letter representing the atom ("O" or "H").

6. Count the electrons you have cut out. You should have 10.

   a. Glue two of these to the oxygen atom, placing them on opposite sides of the dashed circle.

   b. Remember that each bond between hydrogen and oxygen is formed by sharing two electrons (refer to Figure 1). At each place where the edge of the hydrogen atom crosses the edge of the oxygen atom, glue two electrons.

   c. Glue the rest of the electrons (you should have four left) to the oxygen atom spacing them evenly around the solid outer circle.

7. Because of the way hydrogen and oxygen bond, each hydrogen atom has a slightly positive charge, and the oxygen atom has a slightly negative charge. Draw a "−" sign on the oxygen atom and a "+" sign on each hydrogen atom. Do you notice that the "+" signs are on one end of the molecule and the "−" sign is on the other end? This roughly gives the molecule opposite charges on either end—similar to a bar magnet's north and south pole. Molecules with their "+" and "−" charges arranged like this are called polar molecules. Water is considered a strongly polar molecule.

   Because water molecules are polar, they "stick" together much like magnets do (see questions 5 and 6). This sticking together is one reason why water requires so much heat to warm up and even more to boil. It is also responsible for some of water's other properties; for example, surface tension which was explored in the last activity.

Questions/Conclusions

1. Name the elements found in a water molecule.

2. What is the ratio of hydrogen atoms to oxygen atoms in a water molecule?

3. A molecule is a combination of atoms that are bonded

together. How are the oxygen and hydrogen atoms of a
water molecule held together?

4. Describe a polar molecule.

5. If one object has a positive charge and one has a negative
charge, what will they tend to do to each other? (Hint: Have
you ever heard the saying "Opposites attract"?)

6. Using your answer to question 5, why do you think water
molecules tend to "stick" together? What are some different
ways that two water molecules could be positioned so that
they stick together?

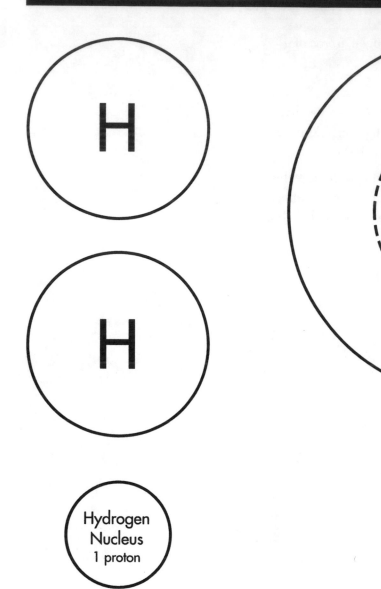

H

H

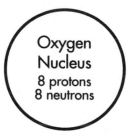

O

Hydrogen
Nucleus
1 proton

Oxygen
Nucleus
8 protons
8 neutrons

Hydrogen
Nucleus
1 proton

Electrons

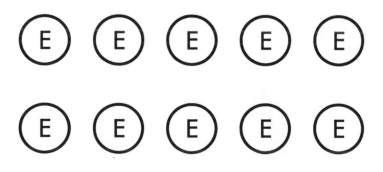

E E E E E

E E E E E

A Sticky Molecule

What is Happening?

Water is a simple molecule, yet no other substance may be more important for life as we know it. A water molecule is made up of only two elements—hydrogen and oxygen. The chemical formula of water is H_2O, meaning that in every water molecule, two hydrogen atoms are bonded to one oxygen atom. Exploring the nature of the bond between hydrogen and oxygen and observing the resulting shape of the water molecule will help students gain insight into water's unique properties.

The atoms in a water molecule bond to each other by sharing electrons—called a covalent bond. The hydrogen and oxygen atoms bond by *sharing* between them hydrogen's lone electron—a hydrogen atom has only one electron—and one electron from oxygen. The sharing, however, is not equal. Oxygen's pull on hydrogen's electron is slightly stronger than hydrogen's own pull. Therefore, the electron spends most of its time closer to oxygen than to hydrogen. Because electrons carry a negative charge, the presence of the extra electron results in a partial negative charge on the oxygen atom. The absence of the electron from hydrogen results in a partial positive charge there. The result is that a molecule of water has poles   like a magnet. This bond—which creates a partial negative charge on one atom and a partial positive charge on the other—is called a *polar covalent bond*. The resulting molecule is called a polar molecule.

In every water molecule, the two hydrogen atoms bond to oxygen in such a way as to make a "mouse head"-shaped molecule, as represented in Figure 2. This shape allows a positively charged hydrogen from one water molecule to move very close to the negatively charged oxygen from another water molecule, as shown in Figure 2. When this happens, the two water molecules are held together by what is referred to as a *hydrogen bond*. It is important to understand that a hydrogen bond only occurs *between* water molecules. The bond that holds hydrogen and oxygen together *within* a water molecule is a different type—a covalent bond.

Although hydrogen bonds are relatively weak, they are numerous enough to influence many of the physical

Materials

For each student:
◊ paper molecule pattern

◊ scissors

◊ glue

◊ crayons or markers (red and blue)

◊ blank paper

Figure 2

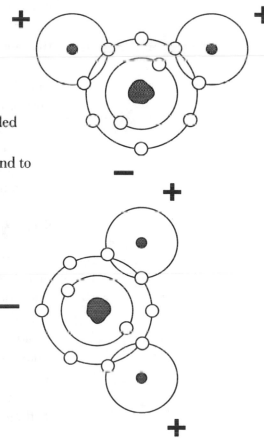

properties of water. For example, it is partially responsible for water's high heat capacity and its high surface tension. This activity demonstrates the polar nature of the bonds *within* a water molecule and the interaction *between* water molecules that results from hydrogen bonding.

Important Points for Students to Understand

◊ The model students construct is a two-dimensional (flat) representation of a three-dimensional molecule.

◊ The polar characteristics of water cause attractive forces between water molecules, which help account for many of the unique properties of water.

Time Management

The construction of the paper model and class modeling may be completed in one class period.

Preparation

Copy one molecule pattern sheet per student; construction paper or card stock is recommended. The hydrogen and oxygen atoms may be copied on different colors of paper or students may color their molecules.

Be sure to include a discussion relating the information in the "What is Happening" section before asking the students to answer question 3 through 6.

Be prepared to point out to students that the "sticking together" they model in response to question 6 results from hydrogen bonds, and that these bonds account for water's high heat capacity, high surface tension, and other properties.

Suggestions for Further Study

Have students build clay models of water molecules. The clay models will give students a more realistic sense of the molecules since they are three dimensional.

Explore the arrangement of water molecules in the solid, liquid, and gaseous states. Use several students' models—if students have made three-dimensional models as suggested above, use them—to show how water molecules are arranged in an ice crystal. Simulate the motion of water molecules in each of its three phases. See Figure 3. You may want your students to

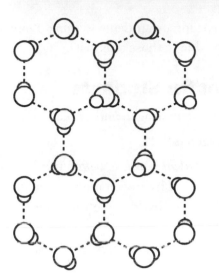
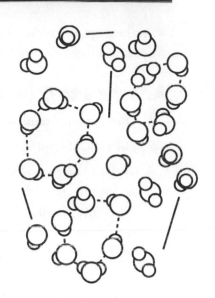
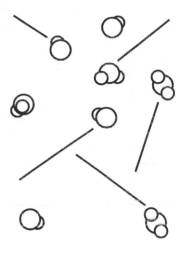

Solid
(many hydrogen bonds)

Liquid
(some hydrogen bonds)

Gas
(no hydrogen bonds)

save their molecules for discussion in Activity 3, *Over and Under—Why Water's Weird.*

Figure 3

Students may wish to make paper models of molecules commonly found in ocean water, such as sodium chloride (NaCl). Students can then demonstrate how water molecules are arranged around both positive and negative ions when a substance is dissolved. See Figure 4. (Also see Reading 1, *Water—The Sum of Its Parts* in the Readings section of this book for a more detailed explanation of molecular structure. Activity 5, *Water—The Universal Solvent* may also be helpful.)

Suggestions for Interdisciplinary Reading and Study

People didn't always explain water's "stickiness," its shape, or how the atoms in a water molecule bond in the same way we do today. Investigate the work of Robert Hooke, Antoni van Leeuwenhoek, and Robert Boyle—to name just a few—and see how their observations and explanations of water and its structure compares with our current explanations. Some of their ideas might seem silly today—like Hooke's belief that bonds were created between molecules by hooks from one molecule intertwining

Figure 4

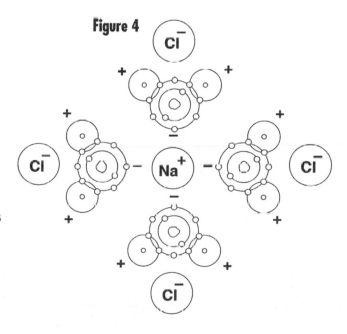

with the loops of another—but encourage students to think about why such a theory seemed reasonable to those scientists.

Answers to Questions for Students

1. The elements found in water are hydrogen and oxygen.

2. The ratio of hydrogen to oxygen is 2:1.

3. *(Be sure you have discussed covalent bonding before you assign this question.)* The atoms of hydrogen and oxygen are covalently bonded. Each of the two hydrogen atoms share one electron with the oxygen atom.

4. When the ends of a molecule are positively and negatively charged, it is said to be polar. Water is a polar molecule because the electron of each hydrogen atom spends more time orbiting the oxygen atom when the two are bonded together, making the electron placement unbalanced.

5. They will tend to attract each other.

6. The positive and negative ends of the water molecules are attracted to one another and cause the molecules to stick together. (NOTE: Students should be encouraged to think of a variety of ways in which two water molecules could orient themselves. Have them work with other students on this question. The correct orientation is shown in Figure 2.)

from *The Rime of the Ancient Mariner*

And now there came both mist and snow,
    And it grew wondrous cold:
And ice, mast high, came floating by,
    As green as emerald.

Samuel Taylor Coleridge

Over and Under—Why Water's Weird

Background

As we've learned, understanding water's molecular structure helps to explain some of its characterisitics. Now let's look at the **density** of water. Density, as you might recall from other science courses or investigations, is the mass of an object divided by its volume. Imagine dropping a small piece of wood in a glass of water. It floats because it is *less* dense than water. Now imagine dropping a piece of metal the same size as the piece of wood— a ball bearing, for example—into the same glass of water. The metal sinks. That's because it is *more* dense than water. (See Figure 1.) This example examines three different substances—wood, metal, and water—but even the same substance can have different densities depending on its temperature.

In general, materials become more dense when they are cooled and become less dense when they are heated. We also know that, generally, when we heat a solid it forms a liquid. (If heated more, this liquid will change into the third state of matter—gas.) It follows then that as most materials are cooled, they become more dense (contract) and sink, while when they are warmed, they become less dense (expand) and rise relative to their surroundings.

In this experiment, we will look at how heating affects the density of different substances.

Procedure

1. Place one of the beakers on the gauze screen on the hot plate. Using the spoon or tongs, put several cubes of the clear gelatin in the uncovered beaker and heat until they just melt.

2. Using a spoon or tongs, carefully place a small cube of colored gelatin into the melted gelatin. Note whether it sinks or floats. ***Caution: Be careful not to splash the melted materials; heated gelatin, shortening, or water can cause severe burns.***

3. Repeat this procedure with the vegetable shortening. Slowly heat several teaspoons of the shortening until just melted. Then carefully place an additional teaspoon of the solid

Objective

The purpose of this activity is to observe the behavior of different susbstances, including water, in their solid and liquid states.

Materials

◊ several cups of ice

◊ several cups of unflavored gelatin cubes—clear and colored

◊ vegetable shortening

◊ teaspoon

◊ spoon or tongs

◊ hot plate with wire gauze screen

◊ three 250 ml beakers

Vocabulary

Density: The mass of an object divided by its volume.

Figure 1

The wood is less dense than water while the metal ball bearing is more dense than water; one floats and the other sinks when placed in a cup of water.

shortening in the newly formed liquid. Note whether it sinks or floats.

4. Repeat this procedure with ice. Slowly heat the ice cubes until just melted. Then carefully place an ice cube in the newly formed water. Note whether it sinks or floats.

Questions/Conclusions

1. Did the solid gelatin float or sink when it was added to the liquid gelatin? Why?

2. Did the solid vegetable shortening float or sink when it was added to the liquid shortening? Why?

3. Did the solid water (ice) float or sink when it was added to the water? Why?

4. Most forms of matter expand when heated and contract when cooled. How is water different?

5. How would life in ponds, lakes, and oceans be affected if frozen water behaved like other frozen substances?

Over and Under—Why Water's Weird

What is Happening?

The solid form of a substance is usually more dense than the liquid form; therefore, the solid form of a substance generally sinks in the liquid form. Water, however, is peculiar. Like other substances, water expands when heated and contracts when cooled. Yet when water is cooled below 4° C it begins to expand rather than contract. At 0° C, it freezes into a solid that is less dense than water (see Figure 2). The result is that ice (solid) takes up more space than an equal mass of water (liquid). It is less dense and therefore floats.

Because water is such a common substance, students may be more surprised to find that the solid form of most other substances *sinks* when placed in the liquid form of that same substance than they are to see that ice floats. Water is also one of the few substances found naturally in all three states. Thus, experience with the solid and liquid states of other substances is limited for most students. This activity demonstrates that while water's properties are more commonly *experienced* and may be viewed as the norm, its properties are actually unique!

The behavior of water in its solid form explains why ice cubes float and why bodies of water freeze from the top down. This phenomenon has significant implications for life on Earth. If ice were more dense than liquid water, water on the top of lakes and the oceans would freeze and sink, allowing additional liquid water on the surface to freeze and sink, resulting in complete freezing of the body of water. If this were to happen, Earth would be significantly different, with much of the water on Earth tied up in a frozen form. Instead, bodies of water freeze on top, insulating lower layers of water from the cold and keeping them from freezing. This allows organisms to live through the winter, somewhat protected from the cold. When warmer weather returns, the ice thaws relatively quickly, returning the body of water to a completely liquid state.

Materials

◊ several cups of ice

◊ several cups of unflavored gelatin cubes—clear and colored

◊ vegetable shortening

◊ teaspoon

◊ spoon or tongs

◊ hot plate with wire gauze screen

◊ three 250 ml beakers

Figure 2:

Freezing Curve of Water

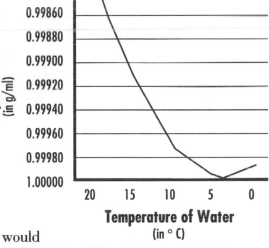

Important Points to Remember

◊ While water is a common substance on Earth, its properties are unique.

◊ The solid form of most substances is more dense than the liquid form. Therefore, the solid form sinks when placed in the liquid form of the substance.

◊ Water is unique in that it is more dense as a liquid than as a solid. Water is most dense at 4°C, while still in its liquid state.

◊ The fact that solid water floats in liquid water has significant implications for life on Earth.

Time Management

This activity takes less than one class period. It also works well as a station activity with other similar activities that deal with water's unique properties.

Preparation

Make the ice cubes and gelatin cubes similar in size and prepare well in advance. Be careful to melt the gelatin, the shortening, and the ice slowly using a medium heat source. If the liquid is too hot the ice and gelatin cubes will melt too quickly for observation. Prepare two to four packets of unflavored gelatin, following the directions on the box. (Note: Unflavored gelatin is used in this activity to emphasize that laboratory materials should not be eaten.) Once the gelatin is dissolved, divide the liquid in half. To one half of the gelatin, add enough food coloring to create a dark color. Leave the other half clear. Chill both batches of gelatin overnight, then cut into cubes. (Note: Paraffin wax may be substituted for the gelatin cubes. However, use caution with hot, liquid wax.)

When selecting a vegetable shortening for this activity, avoid using Crisco or other shortenings that have had air whipped into them during their preparation. Be sure to test the shortening you have chosen prior to the start of the class to be sure you achieve the desired results.

Review student safety procedures concerning working with a hot plate and heated materials. Goggles would be a good precaution. For some classes you may want to use this activity as a demonstration.

Suggestions for Further Study

Use a sample of partially frozen water, a newly formed ice cube with a liquid center, to test its relative position in a beaker of water. Does it float or does it sink?

Have students think of other substances they have observed in which the solid form sinks in the liquid form. Have them think of other substances to test. (Note, however, that many substances are water-based.)

The fact that water expands when it freezes also has important ramifications in geology and the rock cycle. Street paving is also affected by water's expansion when frozen. (What causes potholes?) Have students investigate these relationships through library research or experimentation.

Have students investigate Earth's frozen water—icebergs, frozen lakes, etc.—and how that affects humans, i.e., shipping and transportation, recreation, etc.

Answers to Questions for Students

1. The solid gelatin sank in the liquid gelatin. Gelatin in the solid form is more dense than in the liquid form, therefore the solid sinks.

2. The solid shortening sank in the liquid shortening. Shortening in the solid form is more dense than in the liquid form, therefore the solid sinks.

3. The ice floats on the surface of the water. Solid water, ice, is less dense than liquid water and therefore rises in the water and floats on the top.

4. Water expands when it is heated and also when it freezes. (Water is most dense at 4°C.)

5. Plants and animals that live in the bottom of a pond or lake would be frozen in the water during the cool winter months.

From the traditional sea song
A Wet Sheet and A Flowing Sea

There's tempest in yon horned moon,
And lightning in yon cloud,
And hark the music, mariners,
The wind is piping loud;
The wind is piping loud, my boys!
The lightning flashes free,
While hollow oak our palace is,
Our heritage the sea.

Unknown

How Water Holds Heat

Background

Imagine that you were to put a cold iron skillet and a pot of cold water on the stove at the same time and turn the burners underneath each of them on high. After a few minutes would you be more willing to touch the skillet or put your hand in the water? Most people would be more willing to put their hand in the water than touch the skillet, because they know that some things heat up more quickly than others. This difference between substances results from a difference in **specific heat**, defined as the amount of energy required to raise the temperature of 1 gram of a substance by 1° Celsius.

A substance with a high specific heat must absorb much energy before its temperature increases, and it must loose a lot of energy before its temperature decreases. In contrast, when a substance has a low specific heat, it takes relatively little energy to change the temperature of the substance.

Earth's surface is mostly covered by water. Areas not covered by water—the continents—are surrounded by water. If land and water had the same specific heat, we would expect the land and surrounding water to always be the same temperature. From experience, we know this is not the case. In this activity, we will investigate the heat capacities of sand and water as a simple model of land and a body of water.

Procedure

1. Using a balance, measure 200 g of sand and place it in one of the polystyrene foam cups.

2. Place 200 ml of water in the other polystyrene foam cup. (Note: You can do this using a scale because 1 ml of water has a mass of 1 g. Thus, we are measuring 200 g of water as well.)

3. Attach the two utility clamps to the ring stand.

4. Suspend the thermometers from the utility clamps with the string. Adjust the height of the clamp so that the bulb of each thermometer is just covered by the sand or water. The bulb of each thermometer should be in the center of the container. See Figure 1. *Caution: The lamp will become hot. Take care when using electricity around water.*

Objective

The purpose of this activity is to compare the specific heat capacities of sand and water.

Materials

- ◊ two thermometers
- ◊ ring stand
- ◊ two utility clamps
- ◊ two 240-ml (8 oz.) polystyrene foam cups
- ◊ sand
- ◊ water
- ◊ string
- ◊ lamp setup with reflector and 200-watt bulb
- ◊ clock or watch with a second hand
- ◊ balance
- ◊ graduated cylinder
- ◊ ruler or meter stick

Vocabulary

Specific heat: The amount of energy required to raise the temperature of 1 gram of a substance 1° Celsius.

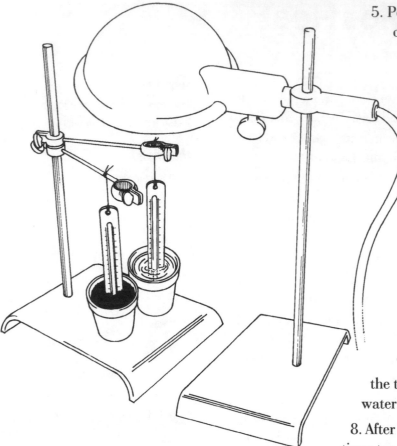

Figure 1

5. Position the lamp 30 cm above the two containers. Be careful to center the lamp so that each container receives an equal amount of light. The setup should look similar to Figure 1.

6. In the data table, record the starting temperatures of both the sand and water. (Note: The temperatures of the sand and water do not necessarily have to be the same.) Also record comments and observations at each reading—such as "water temp. constant for last four min." or "water temp. decreasing 1° C per minute."

7. Turn on the lamp and record in the table the temperature of the sand and water EVERY minute for 15 minutes.

8. After 15 minutes turn the lamp off. Continue to record the temperature of both the sand and water every minute for another 15 minutes.

9. Using the data from the table, plot the changes in the temperature of the sand and water over the 30-minute period. Use the graph provided.

Questions/Conclusions

1. What was the temperature change of the sand and the water after 15 minutes?

2. In which substance did you observe the faster increase in temperature?

3. How many degrees did each substance cool in 15 minutes?

4. Which substance retained its heat energy and cooled more slowly?

5. If the sand and the water received the same amount of light energy, why were there differences in the heating and cooling rates?

6. Describe how differences in the rates of heating and cooling of water and sand could cause the climate along the sea coast to differ from that further inland.

	Time (min)	Sand Temperature	Water Temperature	Comments
Lamp on	1			
	2			
	3			
	4			
	5			
	6			
	7			
	8			
	9			
	10			
	11			
	12			
	13			
	14			
	15			
Lamp off	16			
	17			
	18			
	19			
	20			
	21			
	22			
	23			
	24			
	25			
	26			
	27			
	28			
	29			
	30			

Change in Temperature of Sand and Water over Time

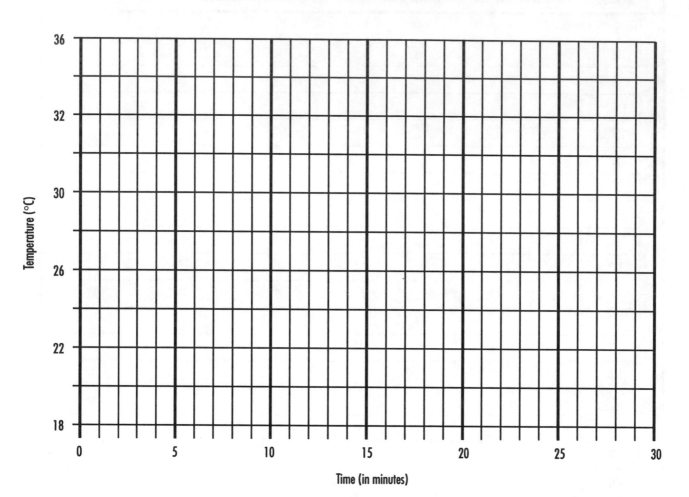

Key

- - - - -   Sand

―――――   Water

How Water Holds Heat

What is Happening?

The specific heat of water is fundamental to the study of oceanography and to developing an understanding of the important role of water in shaping Earth. Specific heat is the amount of heat energy (measured in calories) needed to raise the temperature of 1 gram of a substance 1° Celsius. If you take a given amount of water and an equal amount of another substance at the same temperature, and heat them equally over time, you find the other substance, with two exceptions, becomes hotter than the water. Water heats up slowly because it has a high specific heat. In fact, apart from the two exceptions—hydrogen gas (3.4 cal/g) and liquid ammonia (1.23 cal/g)—water (1.0 cal/g) has the highest specific heat of all substances. (Sand is not a specific chemical compound but rather a mixture, so its heat capacity varies. As a point of reference, the heat capacity of sandstone is approximately 0.26 cal/g. This provides a rough comparison between sand and water.)

In fact, if faced with touching an iron skillet or putting your hand in a pot of water after each had received the same amount of heating as discussed in the student Background section, pick the water. Iron, with its specific heat capacity of 0.113 cal/g, will heat much more rapidly than water. However, instruct your students not to try this.

The specific heat of a substance depends on the size and weight of its atoms. The larger the atoms are, the less heat required to raise its temperature. Specific heat is inversely proportional to a substance's atomic weight.

Due to the high specific heat of water, lakes and oceans change temperature very gradually when heated by the sun or cooled by the atmosphere. Therefore, large bodies of water act as buffers to adjacent land masses and greatly affect climate. Maritime regions exhibit moderate climates, i.e., annual temperature ranges are smaller than inland areas.

This activity explores the concepts of specific heat by comparing the heating and cooling rates of sand and water.

Important Points for Students to Understand

◊ Different materials absorb and release heat at different rates.

Materials

◊ two thermometers

◊ ring stand

◊ two utility clamps

◊ two 240-ml (8 oz.) polystyrene foam cups

◊ sand

◊ water

◊ string

◊ lamp setup with reflector and 200-watt bulb

◊ clock or watch with a second hand

◊ balance

◊ graduated cylinder

◊ ruler or meter stick

◊ The specific heat of water is high. Water's specific heat capacity is responsible for the ocean's moderating effect on the climate of coastal regions.

Time Management

This activity can be completed in one class period. Allow 10 to 15 minutes to distribute and set up materials. The experiment itself requires 30 continuous minutes. Remember to allow time for students to disassemble materials. Make sure each class starts with fresh water and sand, as complete cooling to room temperature may not be achieved between class periods. It may be necessary to extend this experiment over two class periods in order to properly introduce and summarize the activity.

Preparation

No special preparations are required for this activity. However, the sand should be thoroughly dry. Any amount of moisture may interfere with experimental results. Also, it is not critical that the starting temperatures of the sand and water be exactly the same, but it does make interpretation of results much easier. You can achieve equal temperatures in the starting materials by storing the sand and water at room temperature overnight before use.

Since specific heat capacity is defined as the amount of energy required to raise 1 g of a substance 1° C, it is important to adhere to the instructions of using equal masses of sand and water rather that simply filling the cups to the same volume. You may want to make your students specifically aware of this point, especially if you spend time during your laboratory activities discussing the control of experimental variables and sources of experimental errors.

Review student safety procedures concerning use of water and electricity, and for handling a hot lamp.

Suggestions for Further Study

An outdoor experiment may be conducted to demonstrate the difference between the heat capacity of sand and water—reinforcing the Sun's role in the Earth's systems. Set two buckets containing equal masses of sand and water outdoors in an open area with full exposure to sunlight. Record the starting temperature of both the sand and the water; continue taking readings throughout the course of the school day. Compare the results obtained using the lamp to those obtained outdoors.

The concept presented in this activity provides a good opportunity to relate oceanography and meteorology. Differential heating and cooling of land masses and oceans or other bodies of water strongly influences weather, climate, and wind formation.

Ask students to experiment during their own visits to lakes or oceans and try to relate their experiences with the ideas in this activity (i.e., walking across hot sand, wading in cold ocean water, etc.)

Have students investigate how large bodies of water can affect growing seasons in surrounding regions. What effects has this phenomenon had on human demographics?

Have students investigate what "lake effect" is. Using newspaper weather maps or the local weather bureau for information, find out why there is more snow in the winter in certain areas around such large bodies of water as the Great Lakes.

Answers to Questions for Students

1. Answers will vary. The temperature change of the water should be less than the temperature change of the sand.

2. The sand increased in temperature faster than the water.

3. Answers will vary. The temperature of the sand decreases faster than the temperature of the water.

4. The water cooled more slowly, its temperature decreased gradually.

5. Water and sand are made up of different molecules and have different specific heats. The specific heat of water is higher than the specific heat of sand.

6. Because water retains heat so well, the temperatures along the coast do not fluctuate as much as those further inland. The ocean moderates coastal climates. (Differential heating and cooling of land and water is also responsible for some of the local wind patterns associated with coastal regions, such as sea breezes.)

Little Things

Little drops of water, little grains of sand,
Make the mighty ocean, and the pleasant land.
So the little minutes, humble though they be,
Make the mighty ages of eternity.

Julia Carney

Water—The Universal Solvent

Background

Water is often called the universal solvent because so many substances will dissolve in it. Why do so many substances dissolve readily in water?

The ability of a substance to dissolve in a liquid, its **solubility**, is dependent on the molecular arrangements of the liquid and the substance. Water is a polar molecule with positive and negative ends. For this reason, other substances with positive and negative surface charges are attracted to the positive (+) and negative (–) ends of water and are therefore kept "in solution."

Salts—one example is NaCl (sodium chloride), also known as table salt—readily dissolve in water because they contain positive and negative ions that are attracted to the water molecules. See Figure 1. Ocean water contains many types of dissolved salts. Some salts are picked up by rain water and melting snow, are carried to the rivers, and eventually end up in the ocean. Others originate within the ocean when substances such as seashells dissolve in the ocean water and when underwater hot springs add or remove chemicals from seawater. Still others make their way to the ocean through the atmosphere. But the ocean does not grow continually saltier. Salts eventually collect and settle out of ocean water.

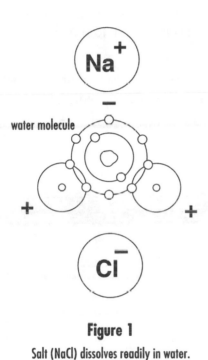

Figure 1

Salt (NaCl) dissolves readily in water.

Objective

The purpose of this activity is to explore the solubility of various substances in water as compared with other liquids.

Materials

◊ water

◊ mineral oil (or baby oil)

◊ isopropyl alcohol (70%)

◊ Epsom salts (Mg_2SO_4)

◊ baking soda ($NaHCO_3$)

◊ table salt (NaCl)

◊ granular sugar

◊ red wax marking pencil

◊ 5 test tubes with rubber stoppers

◊ test tube rack

◊ small spoon (1/8 teaspoon will work)

◊ graduated cylinder

◊ black construction paper

Vocabulary

Solubility: The amount of a substance that will dissolve in a given amount of liquid.

Procedure

1. Number the five test tubes with a pencil or red wax marking pencil. Pour 10 ml of water into the first four test tubes.

2. Measure 1/8 teaspoon of table salt and place it in test tube #5. This test tube will serve as a guide, showing how much solid material you started with.

3. Measure 1/8 teaspoon of table salt, and place the sample into the first test tube. Do NOT stir or shake the test tube. Watch the test tube for one minute. Record your observations in Table 1 estimating the amount of substance that has been dissolved by comparing it with the control test tube (#5) and any other comments about the mixture (clear, cloudy, etc.)

4. Cover the opening of the test tube with the rubber stopper, and shake up and down vigorously ten times. Look at the solution and record your observations. Place a piece of black construction paper behind the test tube to help in determining how much of the material has dissolved. Compare it with test tube #5.

5. Repeat Step 4 until all the salt has been dissolved or until the particles will clearly not dissolve (50 shakes maximum). Record your observations after each set of 10 shakes.

TABLE 1: SOLUBILITY IN WATER

Test Tube	Substance	After 1 min.	10 shakes	20 shakes
#1	NaCl			
#2	Sugar			
#3	Baking Soda			
#4	Epsom Salts			
Test Tube	Substance	30 shakes	40 shakes	50 shakes
#1	NaCl			
#2	Sugar			
#3	Baking Soda			
#4	Epsom Salts			

TABLE 2: SOLUBILITY IN ISOPROPYL ALCOHOL

Test Tube	Substance	After 1 min.	10 shakes	20 shakes
#1	NaCl			
#2	Sugar			
#3	Baking Soda			
#4	Epsom Salts			
Test Tube	Substance	30 shakes	40 shakes	50 shakes
#1	NaCl			
#2	Sugar			
#3	Baking Soda			
#4	Epsom Salts			

TABLE 3: SOLUBILITY IN MINERAL OIL

Test Tube	Substance	After 1 min.	10 shakes	20 shakes
#1	NaCl			
#2	Sugar			
#3	Baking Soda			
#4	Epsom Salts			
Test Tube	Substance	30 shakes	40 shakes	50 shakes
#1	NaCl			
#2	Sugar			
#3	Baking Soda			
#4	Epsom Salts			

6. Use the remaining three test tubes to repeat Steps 3 through 5 for each of the following substances: granular sugar, baking soda and Epsom salts. Record your observations in the data table provided.

7. Repeat the procedure using isopropyl alcohol instead of water in the four test tubes. Record your observations from each trial in Data Table 2.

8. Repeat the procedure using mineral oil instead of isopropyl alcohol in the four test tubes. Record your observations from each trial in Data Table 3.

Questions/Conclusions

1. Which substance, or solute, dissolved fastest in water?

2. Which liquid, or solvent, did not dissolve any substance?

3. Explain why salts readily dissolve in water.

4. Why does shaking increase the amount of substance dissolved?

5. What are some common materials that we dissolve in water?

6. Why is water called the universal solvent?

7. What happens to materials in water that are not dissolved?

Water—The Universal Solvent

What is Happening?

As discussed in previous activities, water is chemically unusual. The fact that it is a polar molecule accounts for all of the properties investigated thus far—surface tension, heat capacity, and the relative densities of its solid and liquid form. Polarity is also responsible for its dissolving ability. Water has the ability to dissolve a variety of substances, and thus water in the ocean contains large amounts dissolved materials. Nearly every element that occurs on Earth is dissolved in seawater.

The solubility of substances, especially those that are ionic, is usually much higher in water than in other solvents. As previously mentioned, the solvent properties of water arise from its polarity. The water molecule contains positive and negative ends. The solute ions are surrounded by the oppositely charged polar ends of the water molecule, causing them to stay apart and not react together to form salt crystals. See Figure 2.

This activity allows students to investigate the solubility of several different substances in three different solvents. Of the three solvents, water has the highest degree of polarity, which accounts for the increased solubility of the substances in water. In addition to the polarity of the solvent molecules, there are other factors which can affect solubility. These include the temperature of the solvent and the degree of mixing of the solute in the solvent. Temperature is not a component of this activity; however, students are asked to investigate the differences in relative solubility due to mixing.

Materials

◊ water

◊ mineral oil (or baby oil)

◊ isopropyl alcohol (70%)

◊ Epsom salts (Mg_2SO_4)

◊ baking soda ($NaHCO_3$)

◊ table salt (NaCl)

◊ granular sugar

◊ red wax marking pencil

◊ 5 test tubes with rubber stoppers

◊ test tube rack

◊ small spoon (1/8 teaspoon will work)

◊ graduated cylinder

◊ black construction paper

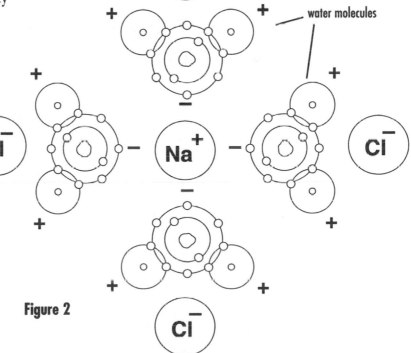

Figure 2

Important Points for Students to Understand

◊ The solubility of a substance in a given solvent is dependent on the molecular arrangement of the solvent and substance involved.

◊ Water is a powerful solvent, dissolving most substances with which it comes in contact. This ability to dissolve many things comes from the fact that water is a polar molecule.

◊ Solubility changes with several factors including mixing and temperature.

◊ River run-off is only one source of the salts found in the ocean. Other sources include seashells, underwater hot springs, and the atmosphere.

Time Management

This activity can be completed in one class period.

Preparation

Depending on class structure and size, teachers may want to measure out small amounts of each solvent and solute for each group of students. Make sure the temperatures of the solvents are the same by storing all solvents for some time at room temperature.

Suggestions for Further Study

Have students expand their experimentation to test the solubility of other substances in water, and to determine how changes in solubility result from changes in temperature.

Have students research the atomic structure of each of the solvents and solutes. Have them look for relationships between structure and solubility. After completing this research, you may ask students to predict the solubility of other substances based on atomic structure.

Investigate the different ways that groundwater or ocean water can become polluted. Include mechanisms that involve dissolving pollutants in water as well as pollutants that do not dissolve in water.

Answers to Questions for Students

1. Answers may vary.

2. Baby oil or mineral oil.

3. Because they contain positive and negative ions which are surrounded by the molecules of water.

4. Without mixing, the substance is at high concentration at the bottom of the container and the remaining material is less likely to dissolve completely. With mixing, the concentration of the substance is uniform throughout the solvent, and the concentration of the substance at the bottom of the container is decreased. The remaining material is more likely to dissolve.

5. Answers will vary, but may include foods (drink mix, powdered milk, instant chocolate, etc.), medicines (Alka-Seltzer, headache powders, etc.), and others.

6. Because so many substances dissolve readily in water.

7. Answers will vary, but may include the following: sink to the bottom of the container or body of water, float on top of the water, or stay in suspension. (Note: This question provides a good opportunity to talk about water pollution. However, do not forget about the pollutants that do dissolve in water.)

Neither Out Far Nor In Deep

The people along the sand
All turn and look one way.
They turn their back on the land.
They look at the sea all day.

As long as it takes to pass
A ship keeps raising its hull;
The wetter ground like glass
Reflects a standing gull.

The land may vary more;
But wherever the truth may be—
The water comes ashore,
And the people look at the sea.

They cannot look out far.
They cannot look out deep.
But when was that ever a bar
To any watch they keep?

Robert Frost

Won't You BB My Hydrometer?

Background

If you have ever gone swimming in an ocean, or better yet, in the Great Salt Lake in Utah, you may have noticed it was easier to float in the ocean or the Great Salt Lake than in a pool or fresh-water lake. Why is this the case? In Activity 5, *Water—The Universal Solvent* we learned that many substances dissolve easily in water. Ocean water and the water in the Great Salt Lake contain large amounts of dissolved salts.

In the first part of this activity, you will construct a hydrometer—a device that allows you to compare the densities of different liquids. The word "hydrometer" means *water* (hydro) *measure* (meter). In this activity you will use your hydrometer to investigate how the addition of salts affects the density of water.

In the second part of this activity, you will use your hydrometer and an egg to monitor the effects of salt in water.

Procedure

Part 1. Constructing a Hydrometer

1. Pick up a tray of materials from your teacher.

2. With a pair of scissors, enlarge the opening of the dropper by cutting off approximately 2.5 cm of its tip. (See Figure 1.) Try to put a BB into the dropper. If the BB fits easily through the opening, go on to step 3. If the BB sticks in the opening, cut off more of the dropper's tip. When the BB rolls into the dropper easily, go on to step 3.

3. Cut a piece of masking tape that is about 6 cm long. Place it on a surface—such as a

Objective

The purpose of this activity is to build and use a hydrometer that will measure the densities of fresh and salt water samples.

Materials for Part 1 (per group)

◊ plastic medicine dropper

◊ sharp scissors

◊ fine-tip permanent marking pen

◊ metric ruler

◊ 20 BBs

◊ 500 ml beaker

◊ masking tape

◊ modeling clay

◊ food coloring

◊ pickling salt

◊ waste container

◊ towels or rags for cleanup

◊ graduated cylinder

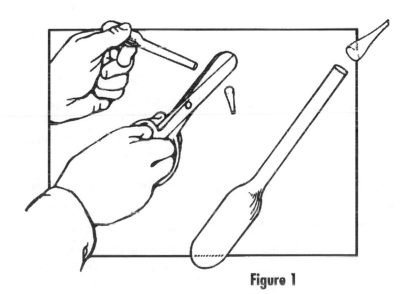

Figure 1

Vocabulary

Hydrometer: A floating instrument that measures the relative densities of liquids.

Density: The mass of a substance per unit of volume.

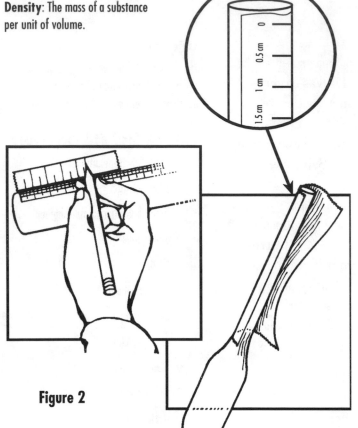

Close-up view

top end of dropper tube

0
0.5 cm
1 cm
1.5 cm

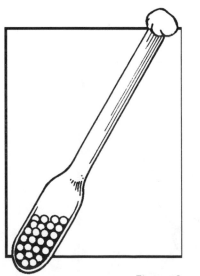

Figure 2

Figure 3

desk—which you will be able to pull the tape back off without ruining the tape. Place the metric ruler next to it and transfer the cm and 1/2 cm markings to the tape. Your piece of tape should look similar to the one in Figure 2. Wrap the tape around the dropper (see Figure 2).

4. Add 400 ml of water to the beaker; add a drop or two of food coloring to the water.

5. Put 12 more BBs into the dropper so that all the BBs rest in the bulb of the dropper.

6. Lower the bulb end of the dropper into the colored water and release it. The bulb of the dropper should remain under water, and the open end should float almost vertically in the water.

7. Add BBs to the dropper one at a time while it is floating in the colored water. Stop adding BBs when only 1 or 2 cm of the dropper tube remains above the water. Remove the dropper from the water.

8. Plug the open end with a small piece of modeling clay as shown in Figure 3.

9. Place the dropper back into the colored water. If it floats so that the surface of the water is anywhere between the 0.5-cm and 2-cm marks, your hydrometer is complete. See Figure 4.

10. If the hydrometer does not float properly, remove the clay plug and the BBs, and go back to Step 5 and proceed as directed. Once your hydrometer floats as described, you can use it to measure how the density of water changes as you add salt.

11. Place the hydrometer, bulb end down, in the colored water used in Part 1. Wait until it stops bobbing up and down.

12. Look at the numbered lines on the tube of the hydrometer. Determine the number of the line that rests on the surface of the colored water. Write this number in the Data Table for

Part 1. If the hydrometer floats so that the water level is between two numbered lines, estimate (using decimal notation) the level where the tube touches the surface of the water. For example, the hydrometer in Figure 4 shows a reading of 1.25 cm.

13. Use the graduated cylinder to measure out 15 ml of pickling salt and pour it into the colored water. Using the hydrometer as a stirring rod, gently stir the water until all the salt has dissolved. Be careful not to disturb the clay plug.

14. Wait until the hydrometer stops bobbing up and down in the water. Determine the number of the line that is just touching the surface of the salty water in the same way you did in Step 13. Write this number in the Data Table for Part 1.

15. Add another 15 ml of dry pickling salt to the water, stir as you did in Step 14, and read the hydrometer when it stops moving. Record your observation for the "colored water + 30 ml salt" on the next line in the Data Table for Part 1.

16. Complete the Data Table for Part 1 by adding salt to the colored water 15 ml at a time as you did in the previous steps. Stir, then make hydrometer readings.

17. Once you have completed all the readings on the different salt solutions, dump the solution into the waste container and return your materials to the teacher.

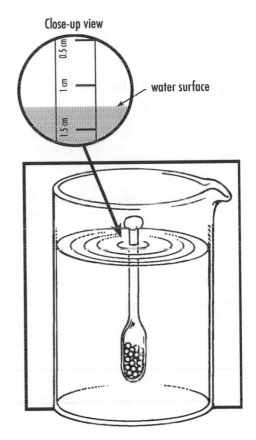

Close-up view

water surface

Figure 4

DATA TABLE FOR PART 1	
Solution Being Tested	**Hydrometer Reading**
colored water	
colored water + 15 ml salt	
colored water + 30 ml salt	
colored water + 45 ml salt	
colored water + 60 ml salt	

Part 2: An Egg in Water

1. Use the graduated cylinder to fill the jar or beaker about 3/4 full with fresh, cool water from the tap. In the Data Table for Part 2, record the amount of water you place in the beaker and, using your hydrometer from Part 1, measure and record the water's density. Remove your hydrometer from the beaker.

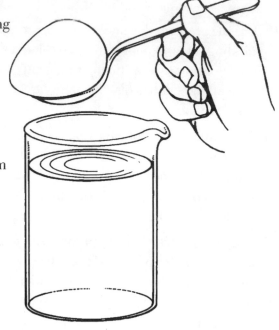

2. Using the soup spoon carefully lower (do not drop!) the egg into the water and observe what happens. Record your observations in the Data Table for Part 2.

3. Remove the egg from the water. Use the measuring spoon to add 10 ml of salt. Use the ruler or straight-edge to level the salt in the spoon before adding it to the water. Add the salt to the water and stir thoroughly with your soup spoon. Measure the density of the solution with the hydrometer. Record the hydrometer reading in the Data Table for Part 2. After you remove the hydrometer, carefully put the egg back into the beaker. Record your observations in the Data Table for Part 2.

4. Repeat step 3 until a change occurs.

Materials for Part 2 (per group)

◊ large jar or beaker

◊ pickling salt

◊ soup spoon (large enough to fit egg on)

◊ 100 ml graduated cylinder

◊ hard-boiled egg

◊ ruler or straight-edge

◊ 5 ml metric measuring spoon

◊ hydrometer (from Part 1)

Questions/Conclusions

1. The more salt that is dissolved in a solution, the _____ [higher/lower] the hydrometer floats.

2. A hydrometer reading of 4.5 cm would mean that _____ [no salt/some salt/lots of salt] is dissolved in a solution.

3. If ocean water gives a reading of 2.5 cm on your hydrometer, what reading might you get if you test fresh water? _____ [1 cm or 2.5 cm or 3.5 cm].

4. Describe how the addition of salt affects some of the properties of water. Is the density of the water increasing or decreasing as salt is added to the water? How do you know whether the change is an increase or a decrease?

5. If an ocean-going ship is loaded with cargo so that 15 meters of the ship rides below the surface of the water, what would you expect to happen when the ship enters the Mississippi River, which is only 16 meters deep at some points?

6. What happened when the egg was placed in the fresh water?

7. Discuss the any differences when the egg was placed in salt water compared to when it was placed in fresh water.

8. How many milliliters of salt did you have to add to make the egg behave differently? Why is it important to use a straight-edge to level the salt?

9. Why do you think things float better in salt water than in fresh water? What does the addition of salt do to the properties of the water?

10. What do you think might happen to the egg if it were left in the beaker for 24 hours? Two days? A week?

11. Compile your results with the results from the rest of the class. Based on the class results, calculate the average amount of salt per 1000 ml of water required to float the egg. Predict how much salt would be needed to float the egg in 20 liters of water and in 300 liters of water. Show your math.

12. Why is it easier to swim in the ocean than in a lake or pool?

DATA TABLE FOR PART 2		Amount of water: _____ ml
Total ml of salt	**Hydrometer reading**	**Result/observations**
0 ml	(initial)	

Won't You BB My Hydrometer?

Materials for Part 1 (per group)

◊ plastic medicine dropper

◊ sharp scissors

◊ fine-tip permanent marking pen

◊ metric ruler

◊ 20 BBs

◊ 500 ml beaker

◊ masking tape

◊ modeling clay

◊ food coloring

◊ pickling salt

◊ waste container

◊ towels or rags for cleanup

◊ graduated cylinder

What is Happening?

The difference between fresh water and ocean water that is most obvious to students is the fact that the ocean is salty. Students may also have some understanding of the fact that the addition of salt to water changes some of the properties of the water. One important change resulting from the addition of dissolved salts is an increase in the density of the water. Water that contains dissolved substances has a greater mass per unit volume (density) than pure water. If 3 g of salt are dissolved in 100 ml of fresh water, the resulting mixture is 3 % heavier than fresh water. Salinity is measured in g per 1000 ml, not per 100 ml, and a special symbol is used: $^0/_{00}$. In other words, a percentage (%) is measured in parts per *hundred*, salinity is measured in parts per *thousand*. The total salt content (in g) dissolved in 1 kg (1000 ml) of ocean water is referred to as salinity and is expressed in parts per thousand ($^0/_{00}$) by weight. Generally, the greater the salinity of ocean water, the greater its density. Open ocean water has an average salinity of about 35 $^0/_{00}$ (or 3.5 percent).

Students may know that it is easier to float in the ocean than in a pool. Some students may have even had the experience of swimming in the Great Salt Lake in Utah, where floating in the water is almost effortless. The more salt dissolved in the water, the less water must be displaced by a floating object. Thus the denser the water, the higher—compared to the water level—an object will float. In this activity, students investigate changes in the properties of water by building and using a hydrometer to measure relative density as increasing amounts of salt are dissolved in the water.

The first part of this activity is an engineering exercise requiring students to read and follow directions in building a hydrometer. This simple device is used by scientists to measure the relative density of solutions. In the second part of this activity students use their hydrometers to track the addition of salts to alter the density of water. At the beginning, the egg has a higher density than the fresh water and thus sinks when placed in the beaker. With the addition of the salt to the water, the water's density increases. When it has a higher density than the egg, the egg floats to the surface. The density of the egg remains constant throughout this activity, while the density of the water increases as salt is added.

With an activity such as this, with a relatively straightfoward procedure, encourage students to follow the principles of good scientific research. Students can control certain variables and are able to make predictions (by recording the amount of water used, by standardizing the amount of salt added, by standardizing the temperature of the water, and by comparing and combining their results with other groups in the classroom). Encourage students to not just follow the directions, but to understand why each step is taken.

Important Points for Students to Understand

◊ The density of water increases with increasing salinity.
◊ The density of any solution, not just a salt solution, can be measured using a calibrated float called a hydrometer.
◊ Hydrometers measure **relative** densities between different liquids. Water is usually the standard for comparison.
◊ Substances of higher density will sink in substances of lower density, and substances of lower density will float in a substance of higher density.

Time Management

Each part of this activity will require one class period. Carefully store the hydrometers overnight.

Preparation for Part 1

No special preparations are required for this activity. It is possible to use a clear plastic straw instead of the medicine dropper simply by plugging both ends with a piece of clay. Be sure to have the students label their hydrometers to avoid confusion on the second day of the activity.

Preparation for Part 2

No special preparations are required for this activity. Pickling salt is recommended because it is pure sodium chloride without additives. Make sure the graduated cylinders the students use are dry when measuring the salt. While students are working through the procedure, draw a table similar to the following Class Data Table (see next page) on the board. Once the students have

Materials for Part 2 (per group)

◊ large jar or beaker
◊ pickling salt
◊ soup spoon (large enough to fit egg on)
◊ 100 ml graduated cylinder
◊ hard boiled egg
◊ ruler or straight-edge
◊ 5 ml metric measuring spoon
◊ hydrometer (from Part 1)

CLASS DATA TABLE FOR PART 2

Group	ml of water	ml of salt to make egg float
class averages		

completed their tables, use their data to compile a class table and averages. Have students compare their results to others in the class and to the class averages.

Suggestions for Further Study

Encourage students to take the hydrometers home and measure the densities of other common household fluids such as dishwashing liquid, cooking oil, real ocean water, water from a nearby creek, etc. Have them keep a record of their findings and compare them with other students from the classroom. You may also want to challenge students to devise a way to use the hydrometer to give readings of salt concentrations in grams per liter; in parts per thousand; in units of their own choosing.

Have students investigate the various uses of hydrometers. For example, hydrometers are used by auto mechanics to test radiator coolant. You may want to invite a local automobile service person to demonstrate the type(s) of hydrometer they use.

You may wish to have students try to float an egg in the middle of the beaker, rather than at the top. This can be accomplished by first adding sufficient salt to float the egg, as in this activity, then slowly adding fresh water to the beaker. Careful addition of the freshwater will result in two layers of water with the egg suspended between them.

Suggestions for Interdisciplinary Reading and Study

As the poem at the beginning of this activity suggests, the sea can be seen as beautiful and captivating. Have students write their own poetry describing a captivating place or experience.

Answers to Questions for Students

1. The more salt that is dissolved in a solution, the **higher** the hydrometer floats.

2. A hydrometer reading of 4.5 cm would mean that **lots of salt** is dissolved in a solution.

3. If ocean water gives a reading of 2.5 cm on your hydrometer, what reading might you get if you test river water? **1 cm**.

4. The addition of salt to the water increases the density of the water, making the hydrometer float higher in the water. The addition of salts to water can also affect the boiling and freezing points of water.

5. The ship would float lower in the fresh water of the Mississippi River than in the salty Gulf of Mexico. This could cause the ship to run aground. Perhaps you could ask the students what they think would happen if the ship was traveling through other liquids. Have students investigate what technologies freighters and other ocean-going ships use to negotiate movement between bodies of fresh and salt water.

6. The egg sank in the fresh water.

7. After a sufficient amount of salt is added to the fresh water, the egg floats.

8. Answers will vary. Because the results are going to be compared to the rest of the class, the leveling insures that everyone is adding the same amount of salt. This precision is an important part of good scientific research.

9. Salt water has a higher density than fresh water, and thus things will more likely float in the salt water than in fresh water. The addition of salt to water increases its density.

10. Answers will vary. Have students try this to test their hypotheses.

11. Answers will vary. Make sure that students make the appropriate conversions between milliliters and liters when calculating the amount of salt required to float an egg in 20 or 300 l of water.

12. The density of the water in the ocean is higher than that in a freshwater lake or pool, thus making it easier for a person to float.

from *Rockall*

Pale ocean rock! that, like a phantom shape,
Or some mysterious spirit's tenement,
Risest amid this weltering waste of waves,
Lonely and desolate, thy spreading base
Is planted in the sea's unmeasured depths,
Where rolls the huge leviathan o'er sands
Glistening with shipwrecked treasures. The strong wind
Flings up thy sides a veil of feathery spray
With sunbeams interwoven, and the hues
Which mingle in the rainbow. From thy top
The sea-bids rise, and sweep with sidelone flight
Downward upon their prey; or, with poised wings,
Skim to the horizon o'er the glittering deep.

Epes Sargent

Ocean Layers

Background

Ocean water is not the same everywhere. In some places the water is colder or deeper than in others. Some parts are more dense or contain differing amounts of dissolved salts than other parts. All these things affect the way ocean water behaves.

Although water is the most abundant substance on Earth's surface, very little of it is pure water. Many elements other than hydrogen and oxygen—the only two elements in pure water—are found in Earth's water. Tap water, for example, contains chemicals used to disinfect the water and to prevent bacterial growth. Ocean water has many other elements—such as dissolved salts—in it.

Water has different properties depending on the environment in which it is found. Water in streams and rivers has few dissolved salts and is called fresh water. Water found in the ocean is called salt water because it has a lot of salt dissolved in it. In areas where rivers flow into oceans—called **estuaries**—the water is **brackish** and has a higher concentration of salts than river water but a lower concentration of salts than the open ocean.

Exploring how these three types of water—salt, brackish, and fresh—mix will help in trying to understand what effects this mixing has on Earth's environment.

Procedure

1. Have a member from your group pick up a tray of materials.

2. Stick the plastic straw into the slice of clay at an angle as shown in Figure 1. (Note: Do not stick the straw all the way through the clay. If the straw comes out through the bottom of the slice of clay, remove the straw and stick it into a different place on the slice.)

3. Test the straw for leaks by filling it with tap water. If the water leaks out of the bottom of the straw, remove the straw and stick it into a different spot on the clay. Test for leaks again if you stick the straw into a different spot.

4. When you are sure that the straw is not leaking, empty the water into your waste container by picking up the entire assembly (the block of clay and the straw) and tipping it so the water drains out of the straw. You

Objective

The purpose of this activity is to investigate what happens when ocean water, brackish water, and river water contact one another.

Materials

For each group:
◊ cafeteria tray

◊ slice of clay 3 cm thick

◊ clear plastic straw (about 10 cm long)

◊ three 250 ml clear plastic cups containing 25 ml each of the colored solutions

◊ 250 ml clear plastic cup (waste container for used solutions)

◊ three medicine droppers or plastic pipettes

◊ one or two sheets of white paper

◊ towels or rags for cleanup

Vocabulary

Brackish water: The type of water found where fresh water and salt water mix. The salt content of the water usually varies and the water is considered neither fresh nor salt. **Estuary:** An partially enclosed area most often near the mouth of a river or other fresh water source, where fresh and salt water mix.

Figure 1

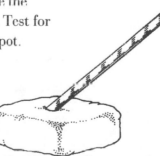

may need to shake the straw several times to get all of the water out, but do not remove the straw from the clay.

5. You know only these two things about the three solutions on your tray:

a. The only difference between the three solutions is their color and the amount of salt they contain.

DATA TABLE: HYPOTHESES AND OBSERVATIONS

Hypothesis 1: _____

_____

Observation: _____

_____

_____

Hypothesis 2: _____

_____

Observation: _____

_____

_____

Hypothesis 3: _____

_____

Observation: _____

_____

_____

Hypothesis 4: _____

_____

Observation: _____

_____

_____

b. If you add small amounts of each solution to the straw in the correct order, you will produce three distinct layers that do not mix.

Your goal is to determine which color corresponds to salt water, which is brackish water, and which is fresh water by adding a bit of each solution to the straw as described in procedure 6. But first write a hypothesis in the data table about the order in which the liquids should be added and explain your hypothesis.

6. Test your hypothesis by adding a small amount of each solution to the straw in the order you described in the guess. Figure 2 illustrates how to do this using a pipette or medicine dropper. Fill about one third of the straw with each solution. Place a white piece of paper behind the straw and observe the colored solutions inside. Record your observations in the data table.

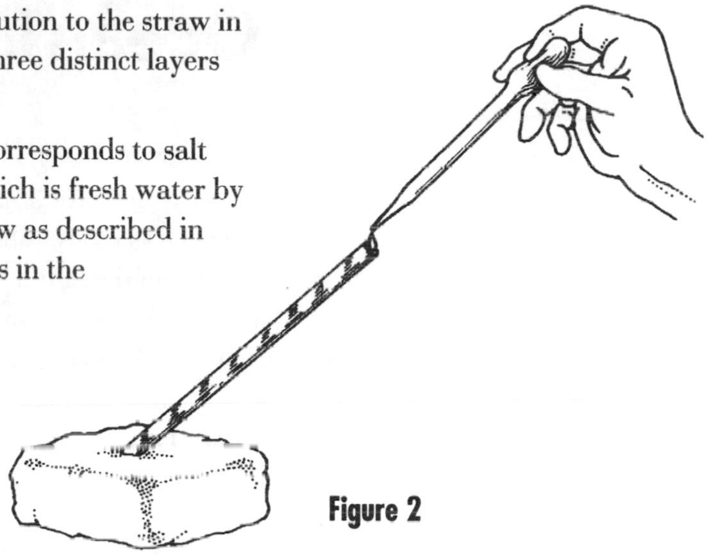

Figure 2

SCI_LINKS.
THE WORLD'S A CLICK AWAY

Topic: ocean floor
Go to: www.scilinks.org
Code: PESO69

7. If you did not see three distinct layers, revise your hypothesis.

8. Empty the contents of the straw and test your revised hypothesis using the same procedure as before. Record your observations.

9. If you did not see three distinct layers, continue revising your hypothesis and testing it. Record new hypotheses and observations in the data table.

Questions/Conclusions

1. What was the order of adding solutions that resulted in three distinct layers?

2. How can you explain this result?

3. Suppose you are at the seacoast in an area where a river runs into a somewhat salty bay before it reaches the ocean. Where do you predict that you would find the **saltiest** water: near the surface of the bay or near its bottom? Explain why.

Ocean Layers

What is Happening?

Water is the most abundant chemical on the surface of Earth. Because so many substances dissolve in it, liquid water is almost never found in a pure form, except in laboratories. Ocean water, for example, is a complex solution of about 96 percent pure water, 3.0 percent NaCl (common salt), and smaller amounts of many other chemicals (see Table 1).

The amount of dissolved salts is not constant from place to place or from time to time throughout the ocean. As a consequence, different regions of water within the ocean can have very different properties. Density of ocean water, for example, is strongly affected by water temperature and amount of dissolved salts. As a result, density varies from place to place. As temperature decreases, density increases until it reaches a maximum at 4°C. Below this temperature, the density begins to decrease. (Remember this from Activity 3, *Over and Under—Why Water's Weird.*) Additionally, as more salts dissolve in water, the water becomes more dense. This trend can continue until the water becomes saturated with salts. Each factor contributes to the variation in ocean water from place to place.

One obvious consequence of differing densities in ocean water is layering, which results in fairly distinct layers of ocean water as depth increases. A less obvious consequence is the formation of density currents. At Earth's poles, the water is both very salty and very cold. These conditions make for very dense water. This water sinks and flows toward the equator along the ocean bottom. Water at the equator flows toward the poles near the surface to replace the sinking water, contributing to global circulation patterns.

Typically, students are aware that different kinds of water have differing amounts of salt dissolved in them. River water, or fresh water, has only a small amount of dissolved salt. Areas where rivers enter the ocean—called estuaries—contain brackish water and have a higher salt content than freshwater rivers. Ocean water contains the highest salt content.

Colored solutions are used in this activity to represent three common types of water: ocean (salt) water, brackish water, and river (fresh) water. The amount of salt dissolved in each is one but by no means the only difference among these three kinds of water. The different concentrations of salt lead to differing

Table 1
Composition of Ocean Water

Element	Percent of Ocean Water
Oxygen	85.7000
Hydrogen	10.8000
Chlorine	1.9000
Sodium	1.0500
Magnesium	0.1350
Sulfur	0.0885
Calcium	0.0400
Potassium	0.0380
Bromine	0.0065
Carbon	0.0028

The following elements are found in seawater at less than 0.001 percent and are listed in decreasing order from Strontium at 0.00081 percent to Radon at 0.0000000000000006 percent: Strontium, Boron, Silicon, Fluorine, Argon, Nitrogen, Lithium, Rubidium, Phosphorus, Iodine, Barium, Aluminum, Iron, Indium, Molybdenum, Zinc, Nickel, Arsenic, Copper, Tin, Uranium, Krypton, Manganese, Vanadium, Titanium, Cesium, Cerium, Antimony, Silver, Yttrium, Cobalt, Neon, Cadmium, Tungsten, Selenium, Germanium, Xeon, Chromium, Thorium, Gallium, Mercury, Lead, Zirconium, Bismuth, Lanthanum, Gold, Niobium, Thallium, Hafnium, Helium, Selenium, Tantalum, Beryllium, Protactinium, Radium, Radon.

densities among these types of water. Fluids having different densities tend to form layers.

Since ocean water contains large amounts of common salt (NaCl) and other dissolved materials, ocean water is more dense than river water. Under certain conditions, river water flowing into brackish estuaries can form a separate layer on top of the more dense ocean water.

Currents, wind conditions, water temperature, and other factors—some of which will be explored in later activities—affect the ways fresh water and ocean water mix or form discrete layers. In this activity, students will discover the effects of amount of dissolved substances on water density and layering. Layering and mixing in estuaries are discussed in more detail in Activity 9, *Estuaries—Where the Rivers Meet the Sea*.

Important Points for Students to Understand

◊ The colored solutions are models for demonstrating how river water, brackish water, and ocean water mix or form layers. Water from natural sources has a greater range of dissolved substances than the solutions in this activity.

◊ When layers form in the ocean, the water containing the most dissolved salt tends to form the bottom layer. The water having the least amount of salt is usually on top, because it is the least dense.

◊ As more solids are dissolved in water, the water becomes more dense.

◊ The density difference between two volumes of water can prevent them from easily mixing. Less dense water will remain on top of more dense water.

Time Management

This activity can be completed in one class period.

Preparation

The pickling salt recommended for the activity produces very clear salt solutions. Other types of salt may produce somewhat cloudy solutions because of various substances added to them. Clear solutions makes it easier to determine whether different samples are mixing or forming layers.

If you use a different kind of salt, test the proportions of salt and water before class to be sure that they give clear separations

Materials

For each group:

◊ cafeteria tray

◊ slice of clay 3 cm thick

◊ clear plastic straw (about 10 cm long)

◊ three 250 ml clear plastic cups containing 25 ml each of the colored solutions

◊ 250 ml clear plastic cup (waste container for used solutions)

◊ three medicine droppers or plastic pipettes

◊ one or two sheets of white paper

◊ towels or rags for cleanup

For the preparation of the colored solutions:

◊ 150 ml of pickling salt (pure NaCl)

◊ 100 ml graduated cylinder

◊ 500 ml graduated cylinder

◊ 600 ml beakers (or jars that will hold more than 500 ml of liquid)

◊ food coloring (three colors: blue, red, green)

◊ stirring rod (or spoon)

◊ marker for labeling beakers

◊ tap water (about 2 liters)

◊ towels or rags for cleanup

of the layers. As a guideline, about the same volume of table salt can be substituted for the pickling salt. If you use coarser salt (such as kosher salt or ice cream salt), you may get better layers if you increase the amount of salt in the "Ocean Water" to about 120 ml in 500 ml water, and in "Brackish Water" to about 40 ml in 500 ml water.

Prepare the colored solutions as follows:

"Ocean Water"	"Brackish Water"	"River Water"
500 ml water	500 ml water	500 ml water
90 ml salt	30 ml salt	no salt
20 drops blue food coloring	20 drops red food coloring	20 drops green food coloring

Stir the "Ocean Water" and the "Brackish Water" until the salt is totally dissolved. Place each solution in a jar or beaker with *no* labeling to indicate the contents.

Alternative procedure:

Have students assist in preparing the colored solutions. The reason for involving students in the preparation of the solutions is so that they may see how the solutions are prepared and understand the differences between them. This reduces the tendency for students to use extraneous facts, such as the color of the water, to explain the events observed.

Suggestions for Further Study

Experimental investigations:

Challenge students to discover the smallest amount of salt that can be dissolved in 500 ml water that will still allow a layer to form between the "Ocean Water" and pure water.

What is the maximum number of discrete layers of colored salt water that can be placed in a straw? In order to answer this, students may wish to experiment with different methods of forming the layers, and test what minimum difference in salt concentration is necessary to form a layer.

Additional investigations:

Students should be encouraged to investigate the different layers of water in the ocean which occur because of density differences. They may also find it interesting to research how long it takes a particle of water to travel in a density current from the poles to the equator (see Activity 8, *The Myth of Davy Jones's Locker*).

Have students explore layering of water in estuaries. The layering produces a dynamic ecosystem with diverse species of animal life.

Ask students to theorize why submarine crews often attempt to conceal their ship's location by keeping it inside a dense layer of ocean water. Library research about submarines may help them find the answers (see Activity 9, *Estuaries—Where the Rivers Meet the Sea*).

Suggestions for Interdisciplinary Reading and Study

Have students read, interpret, and discuss the poem "Rockall" at the beginning of this activity. (Use this suggestion, with some alteration, on other poems and quotes in this book.) Students may work in pairs or groups of four to complete the following.

Make a list of at least five descriptive phrases from the poem, with at least one from each verse. Have students interpret the phrases in their own words, using a dictionary if necessary, and keeping the context of the poem in mind.

Have students explain how the author uses words to "paint" a picture, why the title is, or is not, an appropriate title for the poem, and how the author feels about life at sea.

Have students select a phrase or line from the poem and write it on the front of a piece of construction paper. Students should then illustrate the line on the reverse side in such a way that another student would be able to recite it. Give students a chance to share their cards with the rest of the class.

Answers to Questions for Students

1. Bottom to top: blue (ocean or salt water); red (brackish water); green (river or fresh water).

2. The blue water has the highest density, the green water has the lowest density, and the density of the red water was between the other two. Liquids of different densities will form layers based on their respective densities.

3. Salt water has a higher density than fresh water and would therefore be found near the bottom of the bay. The fresh water from the river would be found near the surface of the bay. Liquids of different densities will form layers based on their respective densities.

From *Apostrophe to the Ocean*

Roll on, thou deep and dark blue Ocean—roll!
Ten thousand fleets sweep over thee in vain;
Man marks the earth with ruin—his control
Stops with the shore—upon the watery plain
The wrecks are all thy deed, nor doth remain
A shadow of man's ravage, save his own,
When, for a moment, like a drop of rain,
He sinks into thy depths with bubbling groan,
Without a grave, unknell'd, uncoffin'd, and unknown.

Lord Byron

The Myth of Davy Jones's Locker

Background

For centuries, sailors believed that bodies buried or lost at sea did not sink to the bottom. They believed that a special depth existed between the surface and the bottom of the ocean where a body would remain suspended. Sailors called this region of the sea "Davy Jones's Locker."

Your teacher has set up a large column in the classroom with vials floating at various levels throughout the column. Study the column carefully and based on what we've learned in previous activities develop some hypotheses that could explain this phenomenon. In this activity, you will create your own model of Davy Jones's Locker.

Procedure

I. Observations

Examine the demonstration column carefully. In the Data Table, record at least five observations about the column that you think are relevant to explaining this phenomenon.

II. Hypothesis

Based on your observations from previous activities you have completed in oceanography, write a hypothesis in the Data Table that you believe explains the phenomenon observed.

III. Prepare your own column

1. Label three beakers and fill them according to the following guidelines:

   Beaker 1 – Ice Cold Ocean Water: Fill the beaker with tap water, adding ice to chill. Stir pickling salt into the beaker until no more will dissolve. (The salinity of ocean water averages 35 g of salt per liter of water. However, the more salt that dissolves in your "ocean water" solution, the easier it will be to set up the water column.)

   Beaker 2 – Cold Tap Water: Fill the beaker with tap water. You may need to add some ice to chill the water slightly, depending upon the temperature of the tap water.

   Beaker 3 – Warm Tap Water: Fill the beaker with hot tap water. This water should be as hot as possible, but not so hot that it will scald someone if spilled.

Objective

The purpose of this activity is to investigate some of the properties of water that could explain the myth of Davy Jones's Locker.

Materials (per group)

◊ three 600 ml beakers

◊ eight small screw top vials (or 8 test tubes and stoppers)

◊ plastic cylinder with end caps (45 cm tall by 4 cm in diameter)

◊ ring stand with clamp to hold cylinder

◊ funnel

◊ 0.5 meter rubber tubing with U-shaped glass tubing in one end

◊ package of BBs

◊ pickling salt

◊ ice

◊ water

◊ red wax marking pencil

2. Add varying numbers of BBs to eight screw top vials until:

- 2 vials sink in Beaker 1 (cold ocean water)— label these vials "A"

- 2 vials sink in Beaker 2 (cold tap water) BUT float in Beaker 1— label these vials "B"

- 2 vials sink in Beaker 3 (warm tap water) BUT float in Beaker 2—label these vials "C"

- 2 vials float in Beaker 3—label these vials "D"

NOTE: It is important to carefully test vials B and C to make sure that they float and sink appropriately. Also, be careful to cap all vials tightly to prevent leaks.

3. Cap the bottom of the tube and place the vials into the empty cylinder. Using a ring stand and clamp, secure the tube so it stands upright (vertical).

Figure 1

4. Carefully pour the water from Beakers 1, 2, and 3 into the cylinder in the following order: Beaker 1 first, Beaker 2 second, and Beaker 3 last. To minimize mixing of the layers, use the device shown in Figure 1—a glass U-shaped tube attached to one end of a 0.5 meter length of rubber tubing and a wide-mouthed funnel.

5. After adding the contents from each beaker, observe what happens. Record your observations in the Data Table.

Questions/Conclusions

The results of this experiment might lead you to believe that Davy Jones's Locker could exist. However, this experiment is entitled *The Myth of Davy Jones's Locker*. As is often the case, experiments conducted within the classroom do not mirror the complexities of the real world, and this is the case with this experiment as well. Davy Jones's Locker does not really exist.

When bodies fall into the ocean and sink, the pressure of the overlying water increases to tremendous levels—far more than could be modeled in the classroom. Increased pressures eventually will crush the body, and it will sink to the bottom.

This activity does, however, highlight some important aspects of how layers form according to density, called density layering. The ocean exhibits a number of layers and this layering is important in understanding estuary formation (which will be examined in Activity 9, *Estuaries—Where the Rivers Meet the Sea*) and deep ocean currents.

1. What are the characteristics of each of the three layers in the column?

2. What do you think is responsible for this layering effect? Explain your response.

3. Evaluate your hypothesis from Part II of the procedure. Were you on the right track?

4. Predict how long these layers would remain distinct if left undisturbed.

5. Evaluate the strengths and weaknesses of this model of oceanic layers.

DATA TABLE : HYPOTHESES AND OBSERVATIONS

Observations: _____

_____

_____

_____

_____

_____

Hypothesis: _____

_____

_____

Observations from Procedure 5: _____

_____

_____

_____

_____

The Myth of Davy Jones's Locker

What is Happening?

Many variables can affect water's density—the amount of dissolved salts (salinity) and temperature are two common examples. Increasing salinity will increase the density of water; increasing temperature will decrease the density of water. In this activity, the layer at the bottom of the cylinder is more dense than the layers above it and therefore stays at the bottom. The top layer is the least dense and floats on the other two. Students should be able to identify the two factors that affect the density of water in this demonstration—salinity and temperature.

These two factors are important in explaining why the ocean is stratified into three separate layers—like the models you and your students have constructed. The surface, or mixed layer contains the warmest and least dense waters in the ocean. The middle layer is a transitional area where water density changes markedly with depth. The bottom, or deep, layer makes up approximately 80 percent of the ocean's volume and contains the coldest, saltiest and, therefore, most dense water in the ocean. While some mixing occurs between them, the layers persist and are well-defined. See Figure 2.

The density differences between the layers and the pull of gravity on water of different densities cause deep ocean currents called thermohaline currents (*thermo* meaning temperature and

Figure 2

Ocean layers

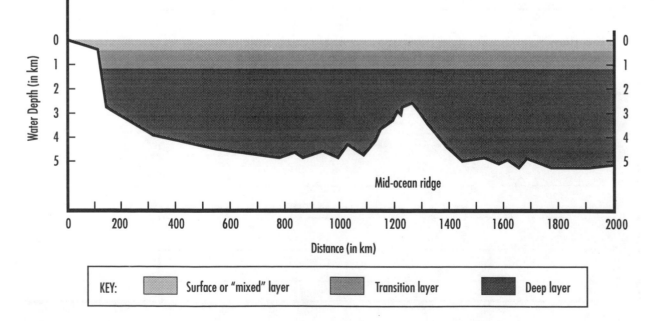

NATIONAL SCIENCE TEACHERS ASSOCIATION

haline meaning salt content). These density-driven currents begin in polar regions. At and near the poles, the water is cold—between 0 and 5°C—and therefore much more dense than water found at lower latitudes. Also, water near the poles is continually being frozen into ice, leaving its salts behind in the unfrozen ocean water. This combination of low temperature and high salinity cause water near the poles to be very dense and sink. Deep ocean currents are established as the dense polar water sinks to the bottom of the ocean and flows toward the equator. These currents move very slowly. Speeds of 1 to 2 cm per second are typical—that is about 70 m per hour. At this rate, it would take more than 15 years for the water in these currents to travel the roughly 10,000 km from the poles to the equator. By contrast, the flow rate of a typical surface current is more than 100 cm per second—at least 50 times the speed of some deep ocean currents.

In this activity, density layering is reinforced and you can introduce students to the concept of deep ocean currents. You may want to refer to this activity during Activity 10, *Current Events in the Ocean* to reinforce the distinction between surface and deep water currents.

Important Points for Students to Understand

◊ Density of liquid water increases with decreasing temperature.
◊ Density of liquid water increases with increasing salinity.
◊ The water in the ocean is layered because of differences in densities of various masses of water.
◊ Differing densities of ocean water result in movement of ocean water.

Time Management

The length of this activity will vary depending on the choice of presentation. Demonstration, observation, experimentation, and discussion of this activity may be completed in one class period.

Preparation

In this activity, students will observe vials suspended at different points in a large demonstration column of water. The demonstration column of water should be set up prior to the start of class. This will take approximately 25 to 30 minutes.

Repetition of this demonstration on a subsequent day, using food coloring to color the various layers, clearly illustrates the relationship between the position of the vials and the different water layers.

The stratified layers of the water column will slowly mix together; however, the cylinder should retain most of its layering for a full day. To help maintain the layers, teachers may wish to place the lower section of the cylinder in a bucket of ice water between class periods and periodically add warm water to the top layer. If left in place over several days, the top layer will lose enough heat that a two-layer column will result—fresh water on the top and salt water on the bottom. Without agitation or mixing, these layers will remain distinct for quite some time.

Materials
(for demonstration set-up)

◊ three buckets (4 liter or 1 gallon size)

◊ eight small screw top vials (or 8 test tubes and stoppers)

◊ plastic cylinder with end caps (180 cm [6'] tall and 6 cm [2.75"] in diameter)

◊ ring stand with clamp to hold cylinder

◊ large funnel

◊ 1 to 2 m length of rubber tubing with U-shaped glass tubing in one end

◊ package of BBs

◊ hot plate and pan or coffee heater

◊ pickling salt (produces a clear brine)

◊ ice

◊ water

◊ red wax marking pencil

◊ optional: food coloring

To prepare the demonstration

Set up the water column as follows (Note: students will go through the same steps to set up their columns):

Label three buckets and fill according to the following guidelines:

1. **Bucket 1** – Ice Cold Ocean Water: Fill the bucket with tap water, adding ice to chill the water. Stir pickling salt into the bucket until no more will dissolve. (The salinity of ocean water averages 35 g of salt per liter of water. However, the more saturated your "ocean water" solution, the easier it will be to set up the water column.)

   Bucket 2 – Cold Tap Water: Fill the bucket with tap water. You may need to add some ice to chill the water slightly, depending upon the temperature of the tap water.

   Bucket 3 – Warm Tap Water: Fill the bucket 2/3 full with tap water. Take the remaining water required to fill the bucket and heat it on a hot plate to just before boiling. Add the heated water to Bucket 3. This produces a warm tap water.

2. Add varying numbers of BBs to eight screw top vials until:

   • 2 vials sink in Bucket 1 (cold ocean water)—label these vials "A"

   • 2 vials sink in Bucket 2 (cold tap water) BUT float in Bucket 1—label these vials "B"

   • 2 vials sink in Bucket 3 (warm tap water) BUT float in Bucket 2—label these vials "C"

   • 2 vials float in Bucket 3—label these vials "D"

NOTE: It is very important to carefully test vials B and C to make sure that they float and sink appropriately. Otherwise, vials may not float in the middle of the column. Also, be careful to cap all vials tightly to prevent leaks.

3. Cap the bottom of the cylinder and place the vials into it. Using a ring stand and clamp, secure the cylinder so it stands upright (vertical).

4. Carefully pour the water from Buckets 1, 2, and 3 into the cylinder in the following order: Bucket 1 first, Bucket 2 second, and Bucket 3 last. To minimize mixing of the layers, use the device shown in Figure 3—a glass U-shaped tube attached to one end of a 1 to 2 m length of rubber tubing and wide-mouthed funnel to the other end of the rubber tubing. NOTE: You may need some assistance from a student or another teacher when you get ready to pour the water into the demonstration tube.

5. The vials should space themselves throughout the column corresponding to water densities—vials D at the surface, vials C at the boundary between the layers of warm and cold tap water, vials B at the boundary between the layers of cold tap water and cold ocean water, and vials A at the bottom of the cylinder.

Materials (per group)

◊ three 600 ml beakers

◊ eight small screw top vials (or 8 test tubes and stoppers)

◊ plastic cylinder with end caps (45 cm [18"] tall by 4 cm [1.5"] in diameter)

◊ ring stand with clamp to hold cylinder

◊ funnel

◊ 0.5 meter rubber tubing with U-shaped glass tubing in one end

◊ package of BBs

◊ pickling salt

◊ ice

◊ water

◊ red wax marking pencil

Suggestions for Further Study

To illustrate the temperature gradient in a column of ocean water, attach a string to a thermometer and lower it into the demonstration tube. Take temperature readings every 5 cm. Record and plot temperature readings on a graph showing the change in temperature with increasing depth.

Add some food coloring to tap water and then freeze the water into ice cubes. Place one of these colored cubes in the water column and observe the results. Ask students to explain what they are seeing.

Suggestions for Interdisciplinary Reading and Study

Have students study and research the myth of Davy Jones's Locker. When and how did this myth arise? Why did it seem reasonable to people at that time?

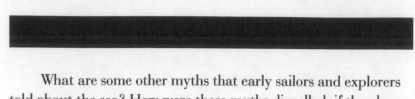

What are some other myths that early sailors and explorers told about the sea? How were these myths dispelled, if they have been? The Bermuda Triangle is a phenomenon that continues to persist today. Have students investigate what scientists or historians says about this and other myths.

Answers to Questions for Students

1. The bottom layer is cold salt water. The middle layer is cool fresh water. The upper layer is warm fresh water. All layers are clear.

2. Density of the water is responsible for the layering observed. Increasing the salt content of water increases its density. Increasing the temperature of the water decreases its density.

3. Answers will vary. Encourage students to conduct an honest evaluation of their hypothesis. Remind students that evaluation and revision are basic common practices in science.

4. Allow the students to experiment with this by leaving one apparatus alone for several days.

5. Answers will vary. Some strengths of this model include: a layered system with little vertical mixing; layers based on relative densities; density differences based on differences in temperature and salinity. Some weaknesses of this model include: ocean layers are not necessarily equal in thickness; except in some estuaries, there are no layers of fresh water in the ocean.

from *The Odyssey*

…While he thus doubted in his mind and heart, a huge wave bore him onward toward the rugged shore. There would his skin have been stripped off and his bones broken, had not the goddess, clear-eyed Athene, given him counsel. Struggling, he grasped the rock with both his hands and clung there, groaning, till the great wave passed. That one he thus escaped, but the back-flowing water struck him again, still struggling, and swept him out to sea. And just as, when a polyp is torn from out its bed, about its suckers clustering pebbles cling, so on the rocks pieces of his skin were stripped from his strong hands. The great wave covered him. Then miserably, before his time, Odysseus would have died, if clear-eyed Athene had not given him ready thought. Rising beyond the waves which thundered on the coast, he swam along outside, eyeing the land, in hopes of finding a sloping shore and harbors off the sea. But when, as he swam, he reached the mouth of a fair-flowing river, there the ground seemed most fit, clear of all stones and sheltered from the breeze. As he felt the river flowing forth, within his heart he prayed.…

Homer

Estuaries—Where the Rivers Meet the Sea

Background

An estuary is a body of water partially enclosed by land that has a connection to a river and an opening to the ocean. Well known examples in the United States are the Chesapeake Bay, Puget Sound, Long Island Sound, and Galveston Bay—there are many others. Lagoons, salt water marshes, and coastal wetlands are estuaries. They are places where fresh water coming from rivers and streams mixes with salty ocean water. The water circulation patterns within an estuary tend to trap the nutrients—washed from the land and downstream through streams and rivers—creating a highly productive area for wildlife. They serve as nursery, spawning, and migration areas for many marine animals.

Many factors affect the mixing of salt and fresh water in an estuary. Large differences in the amount of salt dissolved in ocean water as compared to river water slow down the mixing process. The height of the tides, speed of the outgoing river current, and weather conditions affect how quickly and thoroughly fresh and salt water mix in an estuary. Scientists sampling estuary water sometimes find that the surface water contains very little salt, but a wedge of very salty water lies on the bottom of the estuary. In other cases, sampling indicates little variation in salinity from top to bottom.

In this activity, you will examine a model of an estuary with distinct fresh and saltwater layers. Pay particular attention to how easily the two layers mix.

Objective

The purpose of this activity is to investigate how water mixes in estuaries.

Materials

◊ clear Pyrex™ glass loaf pan

◊ 500 ml "Ocean Water"—the blue solution

◊ 500 ml "River Water"—the yellow solution

◊ pencil (or pen)

◊ 20 cm masking tape

◊ spoon

◊ waste container for used solutions

◊ blank paper

◊ towels or rags for cleanup

Topic: estuaries
Go to: www.scilinks.org
Code: PES085

Procedure

1. Using two pieces of masking tape, each about 10 cm long, tape the pencil to the lab table so that it cannot roll.

2. Place the loaf pan on top of the pencil so that one end of the pan is higher than the other. See Figure 1.

3. Pour the Ocean Water into the loaf pan. Allow the water to become still before continuing. See Figure 2 (next page).

Figure 1

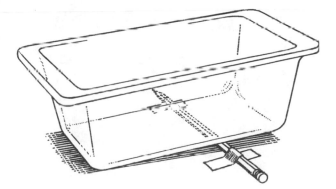

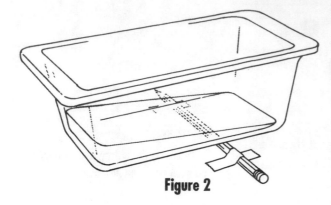

Figure 2

4. Hold the spoon so that it touches the inside of the raised end of the pan. Slowly and carefully pour the River Water down the handle of the spoon so that it trickles into the upper end of the pan without splashing or disturbing the ocean water. See Figure 3. (NOTE: If the handles of the spoons are not suitable for use as described, then very gently pour the River Water into the raised end of the pan. It is important to cause little or no disturbance to the water already in the pan.)

5. Sketch and label—by color—the water layers you see in the pan. Describe how the Ocean Water and River Water mixed, formed mixed layers, or stayed in separate layers as you poured in the River Water.

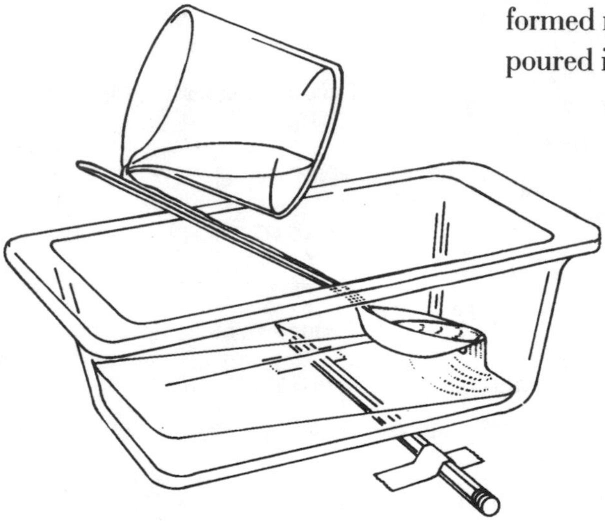

6. Using the pencil as a pivot point, rock the pan back and forth three times. Observe what happens. See Figure 4. This motion is more violent than what occurs in a real estuary, but simulates the mixing created by wind blowing over the surface of estuaries, or by tidal currents flowing into and out of estuaries.

7. Slowly stir the layers of water using the spoon. Observe what happens as you stir. What happens to the layers?

Figure 3

Questions/Conclusions

1. Based on your observations in this activity, describe how the water in a real estuary might:
   a. form separate layers of fresh and salty water.
   b. become a "well-mixed estuary," having no separate layers.

2. Based on your observations from this activity and others you may have completed, what is responsible for the layering effect observed?

Figure 4

3. In what other areas of oceanography, or other areas of Earth science, does the layering of materials play an important role in the characteristics of our world?

4. What are the strengths and weaknesses of this experiment as a scientific model for investigating the structure of an estuary?

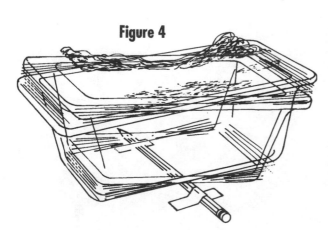

Estuaries—Where the Rivers Meet the Sea

What is Happening?

This activity demonstrates a phenomenon found in many estuaries: the overlaying of dense, salty ocean water by less salty water flowing into the estuary from streams and rivers. In some estuaries, the salty ocean water may form a well-defined, wedge-shaped layer on the bottom. This is commonly referred to as a salt wedge. See Figure 5. The salt wedge remains distinct from the fresh water in this type of estuary.

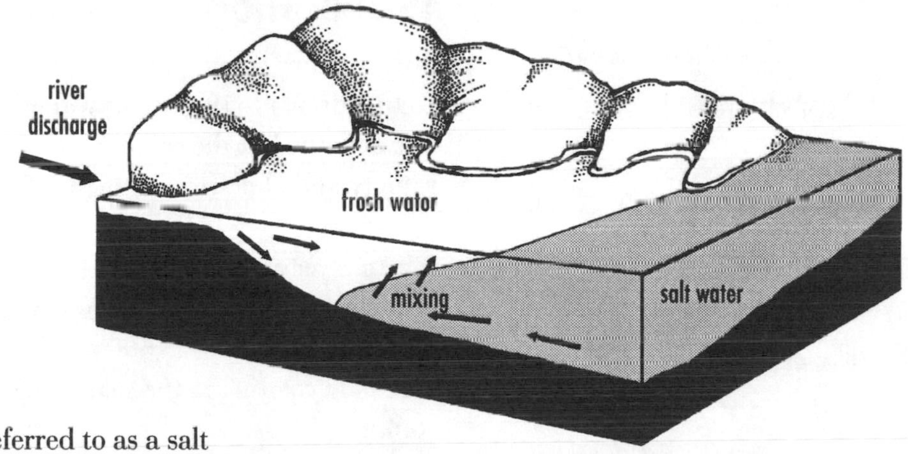

Figure 5

A typical salt-wedge estuary

The results that students obtain for this activity will most often resemble the conditions found in a partially mixed estuary. In this type of estuary, the ocean and river layers are not sharply defined, but there is a definite difference in salinity between the top and bottom layers.

In some estuaries, the conditions are such that the entire water column has a uniform salt content from top to the bottom. In well-mixed estuaries such as this, the salinity is less than that of the ocean, but greater than that of the rivers leading into the estuary.

Important Points for Students to Understand

◊ The saltwater from the ocean and the fresh water from rivers do not mix immediately in any estuary. There is always a tendency for the water to form layers, but in many estuaries, the currents, river flow, and tides mix the water very quickly.

◊ Estuaries are dynamic environments that can change rapidly as a result of ocean storms, heavy rainfall inland in areas that drain into the estuary, or strong wind shifts. In addition to these unpredictable environmental factors, the water in estuaries is influenced by daily tide changes.

Materials

◊ clear Pyrex™ glass loaf pans

◊ "Ocean Water" (see preparation instructions)

◊ "River Water" (see preparation instructions)

◊ pickling salt

◊ food coloring (blue and yellow)

◊ graduated cylinder

◊ pencil (or pen)

◊ masking tape

◊ spoon

◊ waste containers for used solutions

◊ towels or rags for cleanup

◊ Estuaries are extremely important breeding areas for fish and other wildlife.

Time Management

This activity can be completed in one class period.

Preparation

"River Water" is simply tap water. Adding food coloring (3 to 5 drops) to the "River Water" is not essential to performing this activity, but the color contrast between yellow water of low salinity and blue water of high salinity makes the results easier for students to describe. Also, when "River" and "Ocean" water mix, a layer of green water results.

The "Ocean Water" is prepared by adding 90 ml of dry pickling salt crystals to 500 ml tap water. Add 10 drops of blue food coloring to the solution. Stir until the salt is completely dissolved.

Suggestions for Further Study

For the same reasons that estuaries are so nutrient-rich, they are also areas that can become heavily polluted. They trap and store not only nutrients from land, but pesticides, oil, and chemicals that are in runoff and stream and river water. Preserving water quality in estuaries is a major concern of many environmental and sporting groups such as the Nature Conservancy, Ducks Unlimited, and the Sierra Club. Investigate the work of these groups and the efforts of government and industry to deal with the pollution of estuaries.

Many states now have "critical area" laws that try to protect areas like estuaries from pollution and over development. Investigate whether your state or local area has any laws designed to protect estuaries.

If your students live near the coast, encourage them to investigate a nearby estuary.

Suggestions for Interdisciplinary Reading and Study

Have students study and research painters who have painted seascapes, marshes, and estuaries. Paintings of waterfowl, many in estuarine areas, might also be researched. Have students compare and contrast their styles, methods, and composition.

Answers to Questions for Students

1a. A steady flow of fresh water into a calm, protected estuary will tend to produce well-defined layers of water of varying salt concentration.

1b. Estuaries influenced by strong tides, high winds, and variable freshwater flow will tend to become well mixed.

2. Answers may vary, but students should be able to recognize that salt water is more dense than fresh water and that this density difference is responsible for the layering effect.

3. Answers will vary. Some possibilities include: layers within the open ocean, Earth itself, the atmosphere, rocks, etc.

4. Answers will vary. Some strengths include: a layered system with some mixing; water containing differing amounts of dissolved salts do not readily mix. Some weaknesses include: no analogue in nature for the rocking of the loaf pan; an oversimplified system, real estuaries are extremely complex.

From the traditional sea song, *Sailing*

The tide is flowing with the gale,
Y'heave ho! my lads, set ev'ry sail;
The harbor bar we soon shall clear;
Farewell, once more, to home so dear,
For when the tempest rages loud and long,
That home shall be our guiding star and song.

Unknown

Current Events in the Ocean

Background

Sailors have known for centuries that **ocean currents** can speed up or slow down a ship. In modern times, scientists have discovered that ocean currents have major effects on weather patterns and on the ecology of the ocean and nearby land masses. One type of current is called a surface current.

The **Gulf Stream**, which flows along the east coast of the United States, and the **California Current**, which flows along the west coast of the United States, are both surface currents. Surface currents, as you might have guessed, flow across the surface of the ocean almost like a river flows across dry land. However, a surface current does not have solid banks like a river to direct its flow. As a result, the direction of a surface current may change when the wind blowing across it shifts, when it encounters warmer or colder water, or when it nears land.

Procedure

1. Bend the straw at the elbow. Write your name on the short end of the straw using the black permanent marker. This will allow you to tell which straw is yours, and will also remind you which end of the straw to blow into. Do not put the end of the straw with your name written on it into your mouth.

2. Using Map 1 on page 94 (which shows the positions of the continents), draw an outline of the eastern coasts of North and South America and the western coasts of Africa and Europe inside the baking pan with the white chalk. Also draw an arrow indicating north inside the pan, but not in the area representing the Atlantic Ocean.

3. Following the chalk pattern, place ridges of modeling clay along the bottom of the pan to create a boundary system to contain the "ocean." Press the clay firmly to the pan and smooth the gaps between the clay and the pan as shown in Figure 1. It is important to create a water-tight seal to prevent "oceanic" leaks.

4. Fill the ocean area (center) of your model with rheoscopic fluid. Wait for the swirling patterns in the solution to settle.

Objective

The purpose of this activity is to study how landforms and wind affect ocean surface currents.

Materials for 3 to 4 students

◊ baking pan painted black inside

◊ white chalk

◊ modeling clay

◊ colored pencils

◊ a plastic drinking straw with a flexible elbow for each student

◊ black permanent marker

◊ 400 ml of rheoscopic fluid

◊ towels or rags for cleanup

Vocabulary

Ocean currents: Regular movement of large amounts of water in the ocean along certain defined paths. In this activity we are dealing with surface currents, but there are other types of ocean currents.

The Gulf Stream: A specific ocean current that flows south to north along the east coast of the United States. It carries warmer water from the Gulf of Mexico into and across the Atlantic Ocean.

The California Current: A specific ocean current that flows north to south along the west coast of the United States. It carries cooler water from the northern areas of the Pacific Ocean south toward the equator.

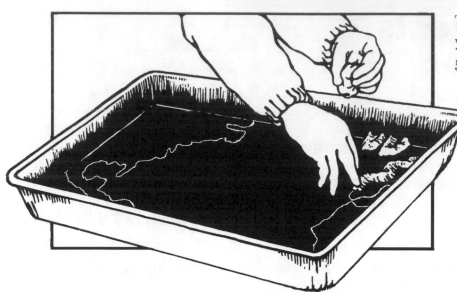

Figure 1

Try not to touch the tray as you do the following steps.

5. For centuries sailors have used the trade winds to cross the Atlantic from Europe and Africa to Central and South America. Simulate these trade winds as follows:

• Hold your straw so that the short end (with your name on it) is parallel to the ocean surface.

• Point your straw from the bulge of Africa westward toward the coast of South America.

• Gently blow through the straw and observe the patterns of ocean currents that the wind produces. See Figure 2.

6. Using one colored pencil, sketch the swirling patterns produced by the wind on Map 1 of the data sheet. With a different colored pencil, use a single arrow to indicate the direction your straw was pointing when you blew. Look at the current that forms in your model. Compare this to the Gulf Stream which runs along the coast of Florida and northward along the east coast of the United States.

7. Alter the direction of the straw, pointing it more northerly or

Figure 2

southerly, and gently blow through the straw. Observe how the current patterns change. On Map 2 of the data sheet sketch the current patterns observed and a single arrow to indicate wind direction. Pay particular attention to what happens as the fluid flows around land masses like Florida.

8. Repeat step 7 at least 3 times (more if time allows) to develop a sense of the relationship between wind direction, landmass configuration, and current patterns. You may wish to put small islands into your ocean and observe their effects on current flow.

9. Repeat Step 5, noticing the region in the middle of the Atlantic Ocean. This region is known as the "Sargasso Sea." Discuss the characteristics of currents in that region.

Questions/Conclusions

1. Describe the relationship between wind and ocean currents. What variables, like wind direction, wind speed, land formations, etc., seem to be related to ocean current patterns?

2. In the late 1700s, Benjamin Franklin described a warm surface current of water running across the Atlantic Ocean from the east coast of North America toward England, what we now know to be the Gulf Stream. Based on the observations you have made with this model and from your knowledge of how water masses containing differing amounts of salt form layers, try to explain why the warm water of the Gulf Stream stays as a distinct current as it moves northward through the colder, saltier Atlantic Ocean waters.

3. The latitude of Edinburgh in Great Britain is approximately that of Moscow in Russia, yet their climates are quite different—Edinburgh experiences relatively mild winters compared with the winters in Moscow. Could this be related to the Gulf Stream? Explain your answer.

4. This setup is a simple model of the Atlantic Ocean and the ocean currents that flow in it. Although scientists use models to help them answer complicated questions like "How do currents form?" the models are limited. Think about how your model is like a real ocean and how it is different. What are some strengths of your model? What are some weaknesses?

SCI LINKS.
THE WORLD'S A CLICK AWAY

Topic: ocean currents
Go to: www.scilinks.org
Code: PESO93

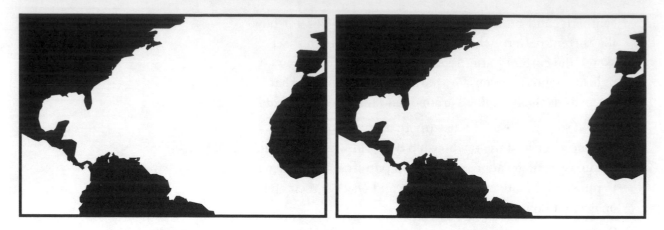

Map 1

Map 2

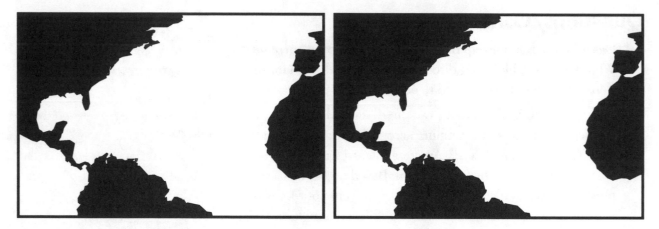

Map 3

Map 4

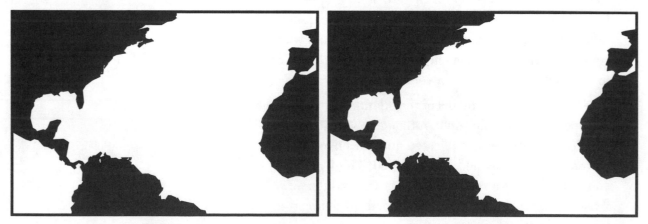

Map 5

Map 6

Current Events in the Ocean

What is Happening?

Currents, large-scale movements of water, occur throughout the ocean—in the surface layers as well as in the subsurface layers. Surface currents are driven primarily by wind. A map showing prevailing wind direction will correlate remarkably well with a map showing surface currents in the major ocean basins. See Figure 4.

This activity allows students to simulate the flow of the Gulf Stream in the Atlantic Ocean and experiment with how land formations and variations in wind direction can affect ocean currents. In actuality, the Gulf Stream flows from the Gulf of Mexico, between Cuba and Florida, and northeastward toward Cape Hatteras, NC. It then curves eastward and flows across the North Atlantic toward Ireland and Great Britain. The friction from permanent wind systems—the trade winds in the south and the prevailing westerlies in the north—appears to be the most important factor in determining the Gulf Stream's motion. It is important, however, to lead students to the understanding that there are many factors that determine current formation in the ocean and this model presents just one of them. Current formation is an extremely complex aspect of oceanography.

Materials for 3 to 4 students

◊ baking pan, 30 cm x 45 cm x 3 cm deep (12" x 18" x 1.5 "), painted black inside

◊ white chalk

◊ modeling clay

◊ colored pencils

◊ a plastic drinking straw with a flexible elbow for each student

◊ black permanent marker

◊ 400 ml of rheoscopic fluid

◊ towels or rags for cleanup

Figure 4

The relationship between global wind patterns and ocean surface currents.

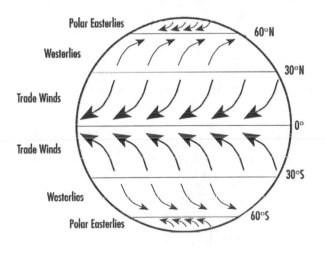

Global wind circulation patterns

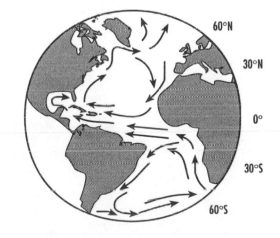

Surface ocean currents in the Atlantic Ocean

Unlike rivers on land, ocean currents may shift their course when influenced by outside factors. These shifts can in turn lead to a variety of changes around the world. In the last 30 years, scientists have been learning more about the cold Pacific Humbolt current—which flows northward along the west coast of South America— and a shift in it that produces a phenomenon known as "El Niño." It has greatly disrupted the South American fishing industry and caused much loss of bird life in nesting sites on small Pacific islands. In the Atlantic Ocean, a shift in the path of the Gulf Stream carried fish that are usually found only in the Caribbean Sea thousands of miles north to the coast of North Carolina.

The reflective properties of the particles suspended in the rheoscopic fluid allow students to see subtle aspects of the currents in a model of the Atlantic Ocean and its adjacent continents. By blowing gently across the model's surface, students generate tiny currents that simulate the huge ocean currents moving millions of cubic meters of water across the ocean.

Important Points for Students to Understand

◊ Surface currents in the ocean are driven primarily by wind.
◊ While surface currents, like the Gulf Stream, are primarily affected by wind direction, other factors—like land formations—influence surface currents.
◊ Scientists are not able to predict when ocean currents will change course. Occasional short-term current shifts occur all over the ocean and can lead to major changes in weather patterns, fish harvests, etc.

Time Management

This activity can be completed in one class period.

Preparation

Spray paint the interior of the trays used for this experiment with a flat black, enamel, rust-proof paint. This type of paint is available at most hardware and paint stores. Allow at least 1 hour for the paint to dry. Other methods of painting or types of paint may be used, but make sure the paint dries with a flat or matte finish, not glossy. A black cafeteria tray may also work, although take care to make sure the clay is built up high enough to prevent the fluid from spilling.

While the rheoscopic fluid is non-toxic, you may wish to require students to wear safety goggles while performing this experiment. As with most foreign materials, should eye contact occur, flush with plenty of water.

Remind students not to share straws.

Rheoscopic fluid is ideal for demonstrating a variety of currents, including convection currents, aerodynamic flow, and much more. It can be reused year after year.(See Appendix B for information on how to obtain rheoscopic fluid.)

Alternative method

As an alternative to rheoscopic fluid, you can substitute water and food coloring.

If you are going to use water and food coloring, you should not paint the aluminum pans black. Paint them instead flat white and the food coloring will show up much better. The procedures will need to be changed as follows: After creating the outline of the continents with the clay, pour water into the area representing the Atlantic Ocean. Put a drop of food coloring in the water where you will begin blowing. Be sure to remind students to watch the direction the food coloring goes. The food coloring will dissipate quickly, so be careful to have students watch closely.

As another alternative, or for demonstration purposes, you may wish to use a hair dryer to create the "wind." Select a dryer that allows you to use a low or no heat setting and a low fan speed. *Use extreme caution if you choose to use an electrical appliance near water.* Use a stand to hold the hair dryer in place, at least 15 cm from the fluid. Students should use appliances only with close adult supervision.

Suggestions for Further Study

Have students make models of other continents and adjacent portions of the ocean. See if you can generate currents in the rheoscopic fluid that simulate known ocean currents. Students can also make islands or jetties along a simulated coastline to study how these can affect longshore current flow.

Students can be encouraged to conduct library research on a number of topics related to the concepts presented here. Some possibilities include: the effects of El Niño in the Pacific Ocean on weather patterns around the world, the ecosystem in the Sargasso Sea and its relationship to ocean surface currents, and the ecological and commercial significance of the Gulf Stream.

Students can investigate explorers' routes and how and why they took those routes. Did currents effect their choices? Look at Columbus' voyage to India, that resulted in his landing in the islands of the Caribbean. What would have happened if ocean currents flowed in the opposite direction? How did he return to Europe and what current or currents aided his return voyage?

Many cultures and peoples, especially cultures that have traditionally depended on the ocean for transport and food—like peoples of the South Pacific—have used knowledge of currents for centuries. Explore some of these cultures—their methods of transport and trade routes.

Suggestions for Interdisciplinary Study

The sea has long been a source of inspiration to poets and songwriters. Have students read the verse at the beginning of this activity and discuss the imagery the author uses. What images of the sea are most vivid in students' minds? What emotions are described in the six lines?

Investigate paintings of the sea. Do the paintings portray the sea as calm and peaceful, or stormy and threatening? When might the sea appear tranquil and unthreatening, or stormy and treacherous? What forces in nature affect the nature of the sea? What are possible dangers of a sea that is stormy? Who is most affected by the "moods" of the sea?

One painter you may want students to investigate is Winslow Homer, a famous American artist of the 19th century. Homer portrayed dramatic moments in the human interaction with the mighty forces of nature found in the ocean.

Homer, born in 1836 in Boston, became an apprentice lithographer at the age of 19. Later he worked as a free-lance illustrator, contributing to *Harper's Weekly* and other publications in which his early works depicted everyday city life and landscapes. He was commissioned by *Harper's Weekly* to illustrate civil war battlefield scenes. Eventually, Homer settled in Tynemouth, England where he became fascinated with the sea. On his return to the United States, Homer lived for years at Prout's Neck, on the Maine coast, where he lived in a cottage and painted seascapes. His favorite motifs included small sailing craft in which the crew were pitted in a struggle against the elements of the sea. His late work in watercolor reflects the influence of Japanese prints in the use of space and sureness of stroke. Not until Homer turned to our relationship to nature and to nature as a subject did he achieve the artistic magnitude for which he is remembered today.

Answers to Questions for Students

1. Wind flow determines current flow in open areas of the ocean and can be instrumental in maintaining flow. Increasing wind speed can increase current speed. Continents or islands can alter the direction of current flow.

2. Because the Gulf Stream is composed of warm, less dense water, it does not mix readily with the cold, dense water flowing down the east coast from the North Atlantic.

3. Yes. The Gulf Stream crosses the Atlantic Ocean as a distinct mass of warm water and moderates the climate of the British Isles.

4. Answers will vary. Strengths might include: it makes a complex system simple, so I can observe what is going on; it lets me change wind patterns to see how different ones affect ocean currents; it creates a good model of how the Gulf Stream flows. Weaknesses might include: it is much smaller than the real thing; factors other than wind cannot be studied; it uses a fluid, not real salt water; it is only one depth, the real ocean varies in depth.

From the traditional sea song,
There's a Hole in the Bottom of the Sea

There's a wing on the fly
on the wart
on the frog
on a bump
on a log
in the hole at the bottom of the sea....

Unknown

Body Waves

What is Happening?

(Please note: This demonstration can be used as an icebreaker or lead-in activity for either Activity 12 or Demonstration 13.) Students who have had the opportunity to go to the ocean or a large lake have an advantage in understanding some of the concepts related to waves. Two such concepts are: (1) the medium in which a wave travels (water in the case of ocean waves) does not move in the same way as the wave itself, and (2) waves transfer, or carry, energy.

If you stand on a pier at the ocean and watch the waves, it is clear that they move toward the shore. This can even be seen in a small pond when a rock is tossed in the middle. While ocean waves travel toward the shore, the water they travel through remains relatively still, moving mostly up and down. People who have floated in the ocean out beyond the breakers know this from first-hand experience.

There are two general types of waves—transverse and longitudinal. Transverse waves, as shown in Figure 1, are the most familiar. Ocean waves are generally classified as transverse waves. In a transverse wave, the medium—water in the case of an ocean wave—**oscillates** perpendicular to the wave direction. In an ocean wave, the *wave* moves toward the shore and the *water* moves up and down, perpendicular to the motion of the wave. Light also travels in transverse waves.

The other type of waves is longitudinal (or compression). Longitudinal waves can be represented by the compression and expansion of a spring, as shown in Figure 2. Sound travels in longitudinal waves. Students may have noticed that very loud sounds can actually hurt their ears. This is because the air in which the sound travels pushes against the eardrum, causing pain. In a longitudinal wave, the medium oscillates parallel to the wave direction. See Figure 2.

Both types of waves can carry and transfer tremendous amounts of energy. Ocean waves provide an excellent example of this. They are capable of knocking people down who are standing in the surf, capsizing huge ships at sea, and transferring massive

Objective

The objective of this demonstration is to investigate the energy of a wave and the motion of the medium in which a wave travels.

Materials

Vocabulary

Oscillate: To move or travel back and forth between two points.

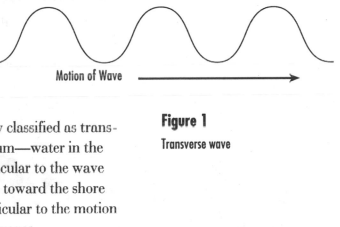

Figure 1

Transverse wave

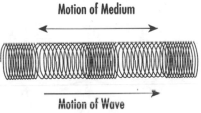

Figure 2

Longitudinal wave

quantities of sand. In many cases, the energy is destructive. However, in some locations around the world, wave energy is actually converted into electricity. This activity allows students to simulate longitudinal and transverse waves and feel the energy these waves transmit.

Procedure

1. Ask nine to twelve students (or whatever number can stand side by side free of obstructions in the room) to participate in the demonstration.

2. Ask the students to stand shoulder to shoulder in a straight line all facing the same direction, preferably toward the rest of the class. The students should be standing "at ease," not bracing themselves or pushing against each other, with each shoulder resting against the shoulder of the student on either side of them. See Figure 3.

3. Stand at one end of the line and give the end student a gentle shove toward the other end of the line. Have the rest of the class make observations. What happened to the students all the way down the line? Where did the energy to make them move come from? How far did the energy travel? Did each student move as far as the energy traveled? What kind of wave was this?
Repeat at least two more times to allow students to make more observations.

Figure 3

4. Ask for a new group of the same number of students. Have them stand side by side all facing the same way and interlock their arms, as illustrated in Figure 4.

Figure 4

5. Stand at one end of the line and pull the end student forward in the direction he or she is facing and then back. Do this several times until every student in the line is moving. Have the rest of the class make observations. Describe the motion of the students in the line—in what direction did they move? What was the direction of the wave? What made them move? Where did the energy come from? Where did the energy go? What kind of wave is this?

Repeat at least once more to allow students to make more observations.

Important Points for Students to Understand

◊ While a wave moves in one direction, there is little movement in the medium through which it moves.
◊ Waves require energy in order to begin.
◊ Waves can transfer energy over long distances.

Time Management

This demonstration takes much less than one class period to do but may take more time depending on the amount of class discussion.

Preparation

The only preparation required is making sure that there is enough open space in the classroom to do the demonstration safely. If the classroom is too crowded, the demonstration could be done outdoors, in a gym or multipurpose room, or in the hall.

Topic: ocean waves
Go to: www.scilinks.org
Code: PESO104

Suggestions for Further Study

Encourage students to study the impact of wave energy. Look at how coastlines erode due to wave action. Beach erosion is a topic around which there is much debate. Should beaches be replenished or should we let nature take its course? Are there ways to protect beaches from waves and erosion? Each of these would make interesting projects.

Challenge students to think of other ways to demonstrate that the water in a wave does not move in the direction of a wave. They may attempt to build a wave tank to show this. Students may suggest that the popular "stadium wave" is a good model of an ocean wave, but in fact it is not. Ask them why. The medium (people) in which a stadium wave travels is not continuous, whereas water is. In a stadium wave, each person has to provide his or her own energy to stand up and sit down. In an ocean wave, the water particles transfer energy from one particle to another. Refer to the activity you just did and the fact that energy transferred from student to student—each student did not have to use their own energy to continue the wave.

Research the changing shape of an island—for example, Nantucket, off the coast of Massachusetts. It is eroding on one end and gaining land on the other end. Explore this phenomenon. Do all islands erode this way? Why or why not? Explore other islands.

Answers to Discussion Questions in Procedure

3. a. They all moved to one side and then back.

 b. The teacher's shove.

 c. All the way to the end of the line.

 d. No.

 e. Longitudinal (or compression).

5. a. Students moved forward and backward.

 b. Toward the end of the line.

 c. The teacher pulling on the end student.

 d. The teacher.

 e. All the way to the end of the line.

 f. Transverse.

Until I Saw the Sea

Until I saw the sea
I did not know
that wind
Could wrinkle water so.

I never knew
that sun
could splinter a whole sea of blue.

Nor did I know before,
as sea breathes in and out
upon a shore.

Lilian Moore

Waves and Wind in a Box

Background

What are some of the ways that you can make waves in water? Have you ever jumped into a pool or lake and done a "cannon-ball"? This makes waves. Have you ever splashed someone else when you were playing in the water? This makes waves, too. Have you ever tried blowing across the surface of water, like in a bathtub or sink? This is still another way to make waves. The tiny ripples you make by blowing across the water are waves just like the much larger waves you see in the ocean.

The largest water waves on Earth are found in the oceans and most ocean waves are created by wind. Far out in the ocean, wind can create waves that are enormous. The largest ever measured accurately was 34 meters high—that is about as high as a 10-story building! Some waves are big enough to capsize huge ships. How can wind make such waves? In this activity, you'll have a chance to find out.

Objective

The objective of this activity is to investigate the relationship between wind and waves.

Materials

◊ one or two large plastic trash bags (preferably white)

◊ 2 kg of sand (optional)

◊ two sturdy cardboard boxes (75 cm x 28 cm x 5 cm or comparable size)

◊ two-speed fan or hair dryer

◊ scissors

◊ packing or duct tape

◊ stopwatch or watch with second hand

Procedure

1. Construct your wave tank by cutting away one end of each box and then joining the two boxes at the open end with tape. See Figure 1.

Figure 1

2. Cut a plastic bag down each side along the crease and unfold it so that it is twice as long. If this is not long enough to cover the bottom and sides of your wave tank, cut another bag and join the two with tape.

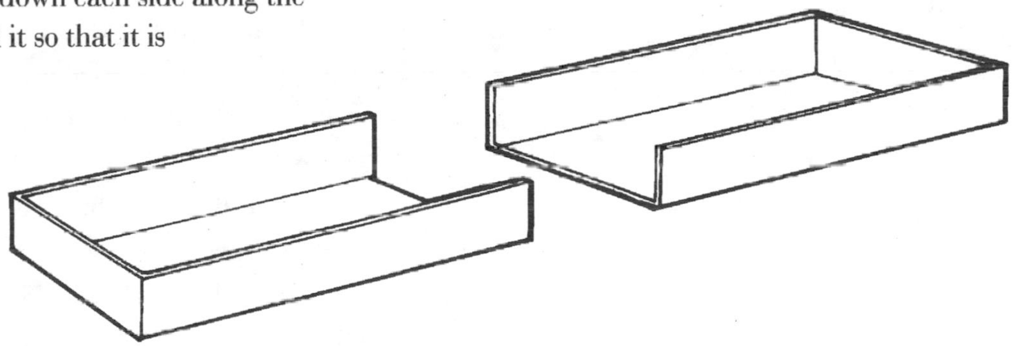

3. With your sand, construct a "beach" at one end of the tank.

4. Fill the tank with water to a depth of about 3 cm.

5. Position the fan or hair dryer so it is aimed down the tank along the surface of the water at about a 45° angle. See Figure 2.

███████████████████████

Part A: Wind speed and wave size

6. You will study how waves are affected by wind. In the Data Table, write a hypothesis that describes how you think the speed of the fan or hair dryer will affect the waves.

7. Turn the fan or hair dryer on "low" for 2 minutes, and record your observations in the Data Table.

8. Turn the fan or hair dryer off and allow the water to become still. Then turn the fan or hair dryer on "high" for 2 minutes, and record your observations in the Data Table.

CAUTION !

Do not allow the fan or hair dryer to touch the water. If either does touch the water, harmful and potentially fatal shock may occur. This warning applies throughout the activity.

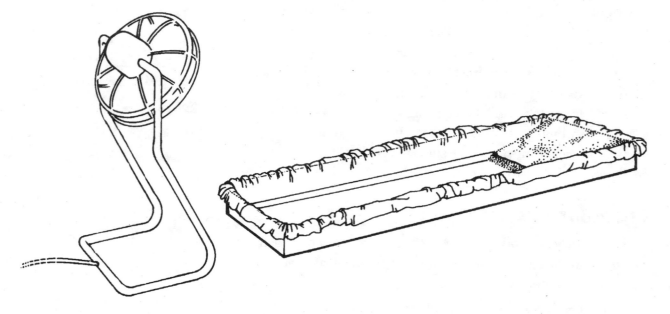

Figure 2

Part B: Length of time wind blows and wave size

9. Write a hypothesis in the Data Table that describes how you think the length of time that the fan or hair dryer blows will affect wave size.

10. Turn the fan or hair dryer on "high" for 5 seconds and record your observations in the Data Table.

11. Turn the fan or hair dryer off and allow the water to become still. Turn the fan or hair dryer on "high" for 10 seconds and record your observations.

Questions/Conclusions

1. In Part A, what happened to the waves when the fan or hair dryer was changed to a higher speed?

2. In Part B, what happened to the size of the waves as time progressed?

3. In Parts A and B, did you notice any differences in the waves from one end of the tank to the other? If so, what were they?

4. From all your observations, what characteristics of wind are important in determining the height of a wave?

	DATA TABLE		
	Hypothesis	**Observations**	
		Fan on low	**Fan on high**
Part A			
		Fan on high, 5 sec.	**Fan on high, 10 sec.**
Part B			

Waves and Wind in a Box

Materials

◊ one or two large plastic trash bags (preferably white)

◊ 2 kg of sand (optional)

◊ two cardboard boxes (75 cm x 28 cm x 5 cm or comparable size)

◊ two-speed fan or hair dryer

◊ scissors

◊ packing or duct tape

◊ stopwatch or watch with second hand

What is Happening?

The most common cause of ocean waves is wind. Waves caused by wind vary in height from only a few centimeters to many meters. The height of waves is determined by three factors related to wind: (1) speed of the wind, (2) the length of time the wind blows in the same direction, and (3) the horizontal distance over which the wind blows—referred to as the "fetch." As each of these increases, the height of waves increases. In a body of water as large as the ocean, these factors can combine to produce waves of enormous height and energy. The largest wave ever measured reliably was 34 meters tall.

The wave tank used in this activity is a very small scale model of what happens in the ocean. Although it accurately represents the role of wind in wave formation, the small size limits the investigation of the third factor, fetch. Fetch can be explored in Demonstration 13, *Tanks A Lot—Activities for a Wave Tank* (also see Suggestions for Further Study).

It is also important to reinforce to students the difference between currents and waves, both of which can be caused by wind. Water molecules in a current move along with the current, resulting in massive amounts of water being moved on a global scale. In a wave, though, water molecules stay in pretty much the same place, and only the energy travels along with the wave. (Refer to Demonstration 11, *Body Waves*.)

Important Points for Students to Understand

◊ Wind causes waves. In fact, most ocean waves are caused by wind.

◊ Wave height is directly related to three factors regarding wind: speed of the wind, length of time wind blows in one direction, and horizontal distance over which wind blows—called "fetch." (Fetch is not addressed in this activity, but it is explored in the Suggestions for Further Study).

◊ The height of waves caused by wind can range from a few centimeters to many meters.

Time Management

Construction of the wave tank and the activity can be done together in one class period, but students may be rushed. Therefore, it may be preferable to spread the two activities over two class periods. Students will probably want to do some additional experimenting with the wave tank.

Preparation

Several of the materials in this activity can be modified. The trash bags do not have to be white, but it does help make the waves easier to observe. Inspect the bags before giving them out to students making sure they have no holes. The sand is optional. However, having a sloped "beach" serves as a dampener to prevent waves from reflecting and interfering with the waves being generated by the fan or hair dryer. To facilitate clean up, the sand can be put in freezer bags and molded into the shape of a beach. This prevents the sand from getting wet. If freezer bags are used, sand can be replaced by sugar, flour, or even dirt.

The cardboard boxes do not have to be of the dimensions listed. Many types can be trimmed to the desired size. A few types of boxes have been found to work well without major modification. One is the very shallow box in which a case of soft drinks is shipped. Another possibility is the lid of a box in which a case of paper is shipped. A third possibility is a sturdy pizza box. With any of these, a wave tank of any length can be made. The length specified in the activity is sufficient for the purposes of this activity, but it would be interesting to construct one that is much longer (sides of a longer tank may need reinforcement).

The fan or hair dryer is not essential. Electricity near water represents a serious safety hazard, as noted in the student section. *If the fan or hair dryer comes in contact with the water, harmful and potentially fatal electric shock may occur.* Another option is to fan the surface of the water with a stiff piece of paper or cardboard, or wind can be generated with a balloon. Very long balloons (one or two meters) are available in most grocery stores. A good source of wind is made available by attaching a straw to the open end and releasing the opening of the balloon while holding onto the balloon itself.

Only one set of wave tanks is needed. The same set can be used for each class.

Topic: tsunamis
Go to: www.scilinks.org
Code: PES0112

Suggestions for Further Study

Because this activity did not have students experiment with fetch—the relationship between wave height and horizontal distance over which the wind blows—have students investigate this relationship by constructing wave tanks of varying lengths. Students can tape three or four boxes together to make longer wave tanks. Be sure to have students compare between the different lengths.

As already mentioned, it is possible to build a much longer wave tank than the one described in the activity. Students can be challenged to devise a way to measure wave height in their wave tank.

Some of the largest and most destructive ocean waves are not formed by wind. They are caused by movements of Earth's crust, such as earthquakes on the ocean floor or huge landslides. These waves are called "tsunamis" or seismic sea waves. Encourage students to investigate the topic of tsunamis—their magnitude, their location, their frequency, etc.

Suggestions for Interdisciplinary Reading and Study

Have students write a poem about waves or the sea using Diamante. See next page.

Study the culture of areas that have long depended on the oceans and seas for their existence. Japan has many accounts of tsunamis—the word is Japanese—that have reached their shores. See how tsunamis have been integrated into their customs, folklore, and even artwork. There is a well-known Japanese painting entitled "The Great Wave Off Kanagawa" which shows a tsunami off the coast of Japan. Find out more about it.

Answers to Questions for Students

1. The waves got bigger.

2. The waves got bigger.

3. Yes. The waves got bigger farther away from the fan, but they were farther apart.

4. Wind speed, andength of time it blows.

Writing Diamante

The Diamante is a poem written in the shape of a diamond. In addition to the shape of the poem, it has two other characteristics:

(1) Its construction is based on three major sentence parts or components. They are *noun, adjective,* and *participle*. These components follow the formula below:

NOUN

ADJECTIVE   ADJECTIVE

PARTICIPLE   PARTICIPLE   PARTICIPLE

NOUN   NOUN   NOUN   NOUN

PARTICIPLE   PARTICIPLE   PARTICIPLE

ADJECTIVE   ADJECTIVE

NOUN

(2) The nouns in the center line (four nouns) are related to the nouns in the first and last lines. The nouns in the first and last lines are antonyms, or opposites, of one another.

Example:

life

green   bright

shining   growing   blooming

heat   motion   sun   food

fading   slowing   dimming

brown   old

death

From *The Forsaken Merman*

Now the great winds shoreward blow;
Now the salt tides seaward flow;
Now the wild white horses play,
Champ and chafe and toss in the spray.

Matthew Arnold

Tanks A Lot—Activities for a Wave Tank

Introduction

Demonstration 11, *Body Waves* and Activity 12, *Waves and Wind in a Box* help make the concepts related to waves concrete for students. They are effective because students generally have firsthand experience with waves—whether at the beach or in a bathtub.

But some concepts related to waves are difficult to demonstrate in the classroom. They require observing water waves on a scale that is not ordinarily feasible for the classroom teacher because the necessary equipment is not available. For example,

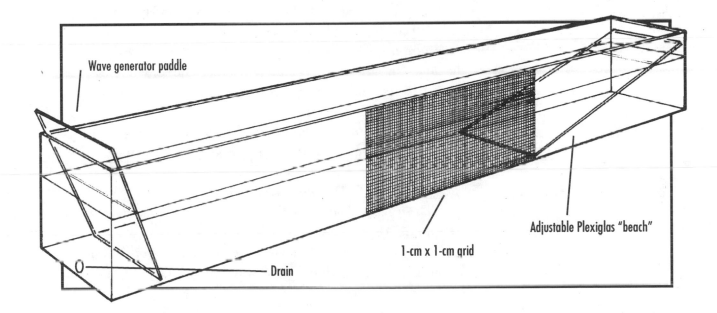

Wave generator paddle

Adjustable Plexiglas "beach"

1-cm x 1-cm grid

Drain

although it is possible to create water waves in a small space, it is difficult to observe, measure, and explore wave height, wavelength, the speed of waves and what factors determine wave speed, and how and why waves break as they approach the shore. The demonstrations presented here relate to these aspects of water waves that are normally difficult for students to observe. They were developed for a large wave tank and designed to be used in a classroom (see Figure 1).

Three demonstrations are presented here: *What's a Wave?*, *Making Waves*, and *Waves: The Inside Story*. A brief "What is Happening?" section relevant to the demonstration is

Figure 1

presented and followed by the procedure. Included in the instructions are suggested observations and questions for class discussion. The demonstrations do not have to be done in the order presented; each can stand alone. The instructions pertaining to filling and draining the tank are written as though the tank is empty at the beginning of each demonstration.

Time Management

The entire set of demonstrations can take as little as one class period or as much as a week depending on how much instruction/discussion the teacher wants to include throughout the course of the demonstrations.

Preparation

The demonstrations described here were developed for a large, clear, Plexiglas wave tank. The tank is approximately 30 cm wide, 30 cm deep, and 270 cm (2.7 m) long. It includes three important features, two of which are attached to the bottom of the tank. One is a wave-generating paddle. The second is a hinged, 90-cm sheet of Plexiglas, used to simulate beaches of varying slope. Finally, a transparent 1-cm-x-1-cm grid is included to attach to one side of the middle 90-cm section. A diagram of the wave tank is shown in Figure 1. These tanks are available commercially (see Appendix B). You can also build one—see Appendix C, *Constructing a Wave Tank*.

The teacher should practice using the wave tank before attempting these demonstrations. Generating smooth waves requires some experience. The teacher should also work out the logistics of filling and draining the wave tank. Filling the tank can be time consuming. It is important to have a hose long enough to go straight from the faucet to the tank. The tank is equipped with a drain valve, as shown in Figure 1. If the tank can be placed so the valve is over a sink, draining will be much simpler.

What's a Wave?

What is Happening?

As discussed and investigated in Demonstration 11, *Body Waves* there are two general types of waves: transverse and longitudinal. Remember that ocean waves are essentially transverse waves (See Figure 2). Also remember that a longitudinal, or compression, wave can be represented as the compression and expansion of a large spring as shown in Figure 3. Sound travels in longitudinal waves.

In both types of waves, it is true that the medium through which the wave moves (water in the case of ocean waves) has little or no net movement. Watching someone floating beyond the breakers demonstrates that while the waves move toward shore, the person bobs up and down as the waves pass. Both types of waves also have crests and troughs. The crest and trough of a transverse wave are illustrated in Figure 2. In a longitudinal wave, a region of compression can be thought of as a crest, and a region of expansion as a trough (Figure 3). Waves also have wave height and wavelength. For transverse waves, wave height is defined as the vertical distance from crest to trough, and wavelength is the distance from crest to crest or trough to trough.

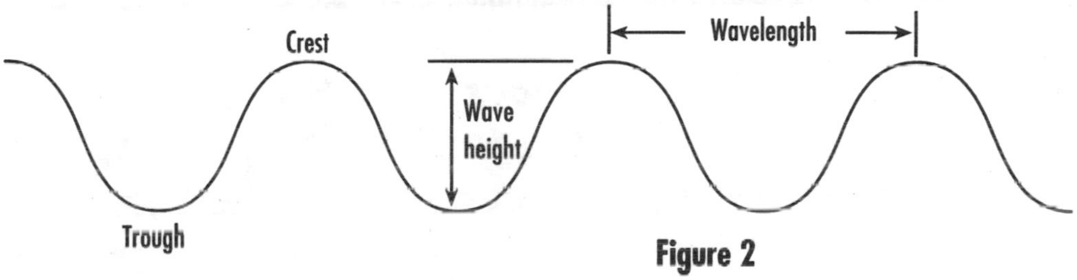

Waves exhibit several properties, two of which are reflection and interference. When a wave strikes an object, it reflects off the object. For example, when sound waves reflect, we may hear an echo. Light waves reflect from objects, and those entering our eyes are the ones we see. The most obvious case of this is light striking a mirror. The reflection of your image in a mirror is similar to an echo of your voice. Water waves reflect, too. This can easily be observed in a swimming pool or bathtub by watching the action of waves as they strike the sides of the pool.

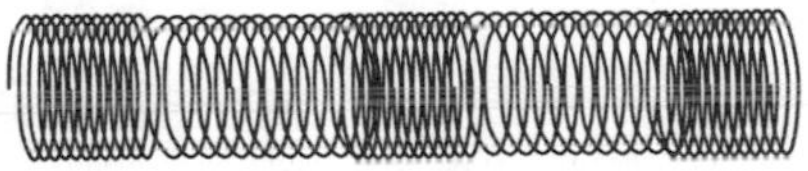

Vocabulary

Crest: The highest point of a wave.
Trough: The lowest point of a wave.

Objective

The objective of this demonstration is to show students selected wave characteristics and properties.

Materials

◊ water

◊ wave tank

Crest

Wavelength

Wave height

Trough

Figure 2

Figure 3

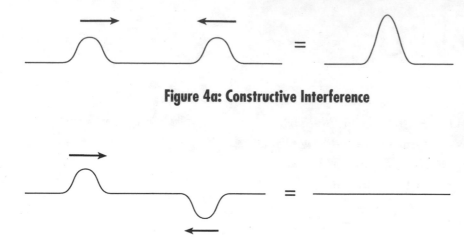

Figure 4a: Constructive Interference

Figure 4b: Destructive Interference

Waves also may interfere with one another. When crests from different waves meet, they will combine, creating a much larger crest (constructive interference) as illustrated in Figure 4a. Sometimes this happens in the ocean and waves large enough to capsize ships are created. If the crest of one wave and the trough of another meet, the two will cancel each other (destructive interference) as illustrated in Figure 4b. Engineers sometimes try to take advantage of this to prevent large ocean waves from coming ashore. They do this by building curved bulkheads near the shore that reflect incoming waves back out into the ocean. The hope is that these will then destructively interfere with larger incoming waves.

Important Points for Students to Understand

◊ Waves have crests and troughs.
◊ Waves reflect when they encounter an obstacle.
◊ Waves can interfere with each other to make larger or smaller waves.

Procedure

Note: For this demonstration, the adjustable piece of Plexiglas, the "beach," (see Figure 1) should be flat on the bottom of the tank.

1. Fill the wave tank with water to a level of about 15 cm.

2. Have a student move the wave-generating paddle back and forth gently in a steady rhythm. This may take some practice.

3. Have students make observations about the waves for at least 3 minutes. Point out crests and troughs.

   NOTE: You can try to measure wave height and wavelength at this point, but it is very difficult to do because of all the interference from reflected waves. It is easier to make these measurements in Demonstration 13C, *Waves: The Inside Story*.

4. Stop generating waves and allow the surface to become almost still.

5. With one forward motion of the wave-generating paddle, make one large wave and watch it travel the length of the tank. Ask students to observe what happens to the wave when it reaches the far end of the tank. Point out that other waves (sound and light, for example) reflect, too.

6. Have a student generate waves again as in Step 2. Ask students to try to pick out two crests moving toward each other from opposite ends of the tank and observe what happens when they meet. It will be easier for them to observe constructive interference if they are close enough to the tank to see the grid clearly.

DEMONSTRATION 13B

Making Waves

What is Happening?

As a wave moves along the surface, the ocean depth affects the wave's speed. This is because a wave actually extends down below the surface of the water. The disturbance of water gradually *decreases* as depth *increases*, and below a certain depth, the wave no longer causes water movement. This depth depends on the wavelength. Generally, the depth below which there is no disturbance is about one-half of the wavelength of the wave. For example, if a wave has a wavelength of 10 m, the water is not disturbed by the wave below a depth of about 5 m. See Figure 5.

The connection between wave speed and water depth will be investigated in this activity. By dividing a specific distance a crest travels by the time taken to travel it, the speed of a wave can be calculated. When the water is shallower than one-half the wavelength, the ocean floor

Objective

The objective of this demonstration is to show the effect of water depth on wave speed.

Materials

◊ water

◊ wave tank

◊ stopwatch

Figure 5

Ocean floor and wave interaction.

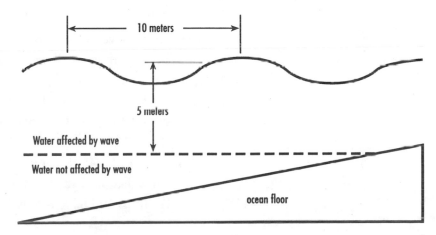

affects the speed of the wave. Friction between the water and the ocean floor slows the wave down. As the water becomes more shallow, the wave slows more and more until it finally reaches the beach and stops. Sometimes, the wave will break as it approaches the beach, an effect discussed in Demonstration 13C, *Waves: The Inside Story*.

Important Points for Students to Understand

◊ Waves extend below the water surface. Therefore waves are affected by underwater factors, such as depth.

◊ In deep water, most waves are not affected by the ocean floor.

◊ As the water becomes more shallow near the shore—less than one-half the wavelength of the wave—the wave speed lessens due to interaction with the ocean floor.

Procedure

1. Fill the wave tank to a depth of 20 cm. The adjustable piece of Plexiglas, the "beach," (see Figure 1) should be flat on the bottom of the tank.

2. Ask for four volunteers—one to generate waves, one to be a timer, one to be an observer, and one to take notes.

3. After the water is almost still, let the wave generator practice sending a single wave to the other end of the tank by moving the wave-generating paddle forward once. At the same time, let the observer practice recognizing when the wave reaches the other end. (Suggestion: Waves can be difficult to see. If a light is directly above the wave tank, the arrival of a wave at the end of the tank can be recognized by the distortion of the light's reflection.)

4. After the volunteers are ready, allow the water to again become almost still. Have the wave generator say "One, Two, Three—Go!", and on "Go", send one wave. On "Go" the timer starts the stopwatch and then stops it when the observer says "Now", indicating that the wave has reached the other end. Record the time on the blackboard.

5. Repeat step 4 at least twice more, allowing the water to become almost still between trials. Calculate the average of the trials and have the recorder write it on the blackboard.

6. Drain the wave tank to 10 cm and repeat steps 4 and 5.

7. Drain the wave tank to 5 cm and repeat steps 4 and 5.

8. Ask students what happened to the time required for the wave to reach the far end of the tank as the depth decreased. Ask them what this implies about the speed of the wave. Ask them why wave speed decreased as the water depth decreased.

DEMONSTRATION 13C

Waves: The Inside Story

What is Happening?

Water in an ocean wave has little or no net motion. Water waves were described as transverse waves, and for most purposes, this description is sufficient. Strictly speaking, water waves are not transverse but are instead "surface waves." Surface waves occur at the boundary between two materials; for example, between water and air (water waves) or between the ground and air (earthquake waves). A surface wave is a combination of a transverse and a longitudinal wave. (See the "What is Happening?" section of Demonstration 13A, *What's a Wave?* for descriptions of these waves.) That is, the particles in these waves move up and down and side to side, making for a nearly circular orbit. The net movement of particles in surface waves is still almost zero.

Things floating in the ocean as a wave comes by display this type of motion. If you were on a raft floating out beyond the breakers, this is what you would notice: As a crest approached, you would move backward (away from the shore) and up, then forward and up until the crest was directly underneath you. At this point you would have completed half your circular orbit. As the crest passed, you would first move forward and down and then down and back until you completed the orbit. This is depicted in Figure 6. The motion is very similar to what you experience on a Ferris wheel but on a much smaller scale.

As an ocean wave approaches the shore, the depth of the water decreases. At a depth of one-half the wavelength, the ocean floor begins to affect the wave. The first effect is a slowing of the wave. As the water becomes even more shallow, the shape of the wave begins to change. This is because of the influence of the ocean floor on the orbital motion of the water particles. As the

Objective

The objective of this demonstration is to show the orbital motion of particles in an ocean wave and two aspects of wave-coast interactions: breakers and sand transport.

Materials

◊ water

◊ wave tank

◊ small pebbles

◊ floating objects (Ping Pong balls, eyedroppers, corks, etc.)

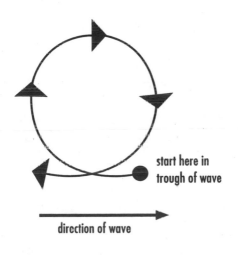

start here in trough of wave

direction of wave

Figure 6

Direction of particle movement

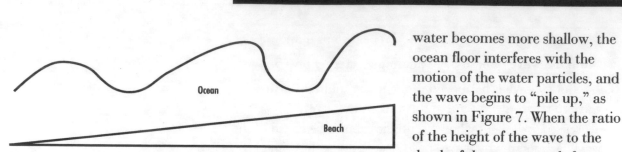

Figure 7

The effect of the ocean floor on the shape of a wave.

water becomes more shallow, the ocean floor interferes with the motion of the water particles, and the wave begins to "pile up," as shown in Figure 7. When the ratio of the height of the wave to the depth of the water equals four-fifths, the wave becomes unstable and breaks. Not all waves break, however. Whether or not they break is determined by several factors. Among them are the speed of the wave, the height of the wave, and the slope of the beach.

Ocean waves carry tremendous amounts of energy. If you have ever been underneath a wave when it broke, you know this to be true. The energy is dissipated in several ways. One of these is in moving sand and other things on the beach. In some cases, beaches are eroded. In other cases, they are built up.

Important Points for Students to Understand

◊ Water particles in a wave are not carried along with the wave. Instead, they have little or no net movement.

◊ The water in a wave moves in a nearly circular pattern.

◊ As water becomes shallower, the ocean floor begins to affect the orbital motion of the water particles and changes the shape of the wave.

◊ As the wave approaches shore, its shape may become unstable enough to collapse, forming a breaker.

Procedure

1. The wave tank should be filled to a depth of 20 cm.

2. Raise the adjustable piece of Plexiglas (see Figure 1) so that it makes a 20° angle with the bottom of the wave tank.

3. Move the wave-generating paddle in a smooth, rhythmic way and have students make observations for at least three minutes. They can either do this aloud or in writing. Ask them what happens as the wave approaches the "beach."

4. Adjust the piece of Plexiglas to increase or decrease the angle and have students make more observations.

   NOTE: The "beach" acts as a wave dampener and prevents most wave reflection. Therefore, this is a good time to attempt measuring wave height and wavelength.

5. Allow the surface of the water to settle and return the beach to a 20° angle. Add several small pebbles to the beach at the water line and continue making waves. (Pebbles are recommended instead of sand because they are easier to remove from the tank.) Ask students to observe what happens to the pebbles. Ask them what they think waves would do to sand based on their observations of the wave tank.

6. While still making waves, ask the students to specify in which direction the waves are traveling. They should say that the waves are traveling toward the "beach."

7. Allow the water to settle and add floating objects to the wave tank close to the side with the grid on it.

8. Generate waves and ask students to observe the motion of the floating object. Is the direction of motion the same as that of the wave?

9. Have students come up and trace with their finger or a water-based pen the path the object travels when a wave goes by. It is essential that the object be close to the grid on the side of the tank for this.

Suggestions for Further Study

The teacher is limited only by his or her imagination in the number of demonstrations and student activities that can be done with the wave tank.

Many schools have video cameras. Students can carefully investigate interference by videotaping the events and viewing them "frame-by-frame."

Many types of wave action are observable in the tank. Students may want to do a more controlled experiment concerning the effect of the floor of the ocean on wave speed. They may also want to try other ways of measuring wave height and wavelength. Wave-coast interactions (particularly sand transport) are a relevant topic. Students can use the wave tank to learn why coastal barrier islands are constantly shifting.

The topic of waves in general holds many possibilities for additional investigations. Encourage students to study the topic of sound and compression waves. Sound waves are used to determine water depth in oceans. Students may be interested to learn how this technique is used and how sound travels in water as compared to air.

The Tide

The tide rises, the tide falls,
The twilight darkens, the curlew calls;
Along the sea-sands damp and brown
The traveller hastens toward the town,
And the tide rises, the tide falls.

Darkness settles on roofs and walls,
But the sea, the sea in darkness calls;
The little waves, their soft, white hands,
Efface the footprints in the sands,
And the tide rises, the tide falls.

The morning breaks; the steeds in their stalls
Stamp and neigh, as the hostler calls;
The day returns but nevermore
Returns the traveller to the shore,
And the tide rises, the tide falls.

Henry Wadsworth Longfellow

Plotting Tidal Curves

Background

Anyone who has been to the coast realizes there is a rhythm to the ocean. Waves crash onto the beach or rocks. The water they carried washes ashore, then retreats. Another wave crashes ashore. The pattern repeats. However, there is another rhythmic phenomenon along the coast—tides.

Tides are predictable changes in sea level that occur at regular intervals. There is a high tide when sea level has risen to its highest point and at low tide it falls. These changes affect humans in many ways. A person who falls asleep on the beach at low tide may wake up drenched if he/she does not move before high tide comes. One of the most important impacts of tides is on ocean shipping. In many locations, ships can only come to shore at high tide. If they come in at low tide, they may become grounded.

Because tides affect us in so many ways, it is important to know when they will occur. As already mentioned, it is possible to predict when tides will occur. By making a graph of the time of the tides and the sea level, you can begin to see the pattern of changes over a period of time. Although this can be done for any location on the coast, the location used in this activity is Ocracoke Island on the Outer Banks of North Carolina.

Objective

The objective of this activity is to plot tide data for a period of one month and draw the tidal curve for this data.

Materials

◊ pencil with eraser

◊ red pen or a bright color crayon

◊ scissors

◊ clear tape

◊ ruler

SCi**LINKS**
THE WORLD'S A CLICK AWAY

Topic: tides
Go to: www.scilinks.org
Code: PES0125

Procedure

1. Locate the sheets titled "June 1991—Ocracoke Island, NC Data Sheet" and "Graphing Charts for Tide Data." Time is given in the military convention (or 24-hour time). Until 1200, or 12:00 noon, the two conventions are the same. After that just subtract 1200 from the military time to get the civilian time. For example, 1430 means that it is 2:30 p.m. On the graphing charts, 6 corresponds to 0600 in the time column of the "June 1991–Ocracoke Island, NC Data Sheet."

2. Plot the data on the graphing chart with your pencil. It is important that you plot the points with the pencil so that if you make mistakes, they can be changed. After you have plotted all the data, go back and check for correctness.

3. With your pencil and ruler, connect all the points in the order they were plotted. Check for correctness.

4. With your pen or crayon, go over the pencil lines. This will make them much easier to see.

5. Cut out the three charts. Along the bottom of the charts, you will notice some circles on days 5, 12, 19, and 27 that represent the phases of the Moon. Be sure not to cut these out when you cut the charts.

6. Tape the three charts together so that they make one continuous chart.

Questions/Conclusions

1. About how much time passes between one high tide and the next high tide? From one low tide to the next low tide?

2. When did the *highest* high tide occur? The *lowest* low tide?

3. When did the *lowest* high tide occur? The *highest* low tide?

4. You are staying at a beach house. At high tide, the water completely covers the part of the beach that is usable. On June 6, you go out on the beach at 10:00 a.m. to read a book. Will you be able to find a dry place to sit down? After an hour the sound of the waves lulls you to sleep. How long can you sleep before you must either wake up or get wet?

5. Ocracoke Island has shallow sand bars all around it that will prevent even small sailboats from coming close to shore at low tide without being grounded. But at high tide, these sand bars do not cause a problem for small boats.
On June 20, the crew of a sailboat wants to get as close as possible to the island without getting stuck on the sand bars. They radio you at the Coast Guard station to ask what time they should come in so as to take advantage of high tide and daylight. What should you tell them? (Give the time in civilian convention.)

Day	Time	Sea Level (feet)	Day	Time	Sea Level (feet)	Day	Time	Sea Level (feet)
1	0432	0.72	11	0047	0.32	21	0102	2.52
	1003	2.72		0615	2.92		0927	0.52
	1625	0.82		1236	0.22		1541	3.12
	2215	3.02		1842	3.72		2231	0.62
2	0510	0.82	12	0141	0.22	22	0358	2.52
	1147	2.72		0710	2.92		1018	0.62
	1707	0.92		1331	0.12		1632	3.12
	2254	2.92		2037	3.72		2326	0.62
3	0547	0.82	13	0233	0.12	23	0450	2.52
	1130	2.72		0805	3.02		1109	0.62
	1757	0.92		1427	0.12		1720	3.12
	2349	2.82		2030	3.72	24	0004	0.62
4	0628	0.82	14	0324	0.02		0539	2.52
	1216	2.72		0900	3.12		1153	0.62
	1850	1.02		1526	0.12		1804	3.12
5	0030	2.72		2123	3.62	25	0059	0.62
	0713	0.82	15	0416	0.12		0623	2.62
	1307	2.82		0956	3.12		1239	0.62
	1950	0.92		1623	0.22		1846	3.22
6	0122	2.72		2217	3.42	26	0139	0.62
	0802	0.72	16	0507	0.12		0705	2.62
	1403	3.02		1051	3.12		1322	0.62
	2051	0.92		1723	0.32		1924	3.22
7	0222	2.62		2313	3.22	27	0218	0.62
	0854	0.72	17	0557	0.22		0746	2.72
	1450	3.12		1149	3.12		1403	0.62
	2154	0.72		1824	0.42		2001	3.22
8	0321	2.72	18	0009	3.02	28	0254	0.62
	0949	0.62		0650	0.32		0825	2.72
	1554	3.32		1548	3.12		1443	0.72
	2255	0.62		1927	0.52		2036	3.12
9	0422	2.72	19	0104	2.82	29	0329	0.62
	1045	0.42		0743	0.42		0902	2.72
	1650	3.52		1345	3.12		1524	0.72
10	2352	0.42		2032	0.62		2112	3.12
	0519	2.82	20	0204	2.62	30	0403	0.62
	1140	0.32		0835	0.52		0940	2.82
	1748	3.62		1444	3.12		1601	0.82
				2135	0.62		2146	3.02

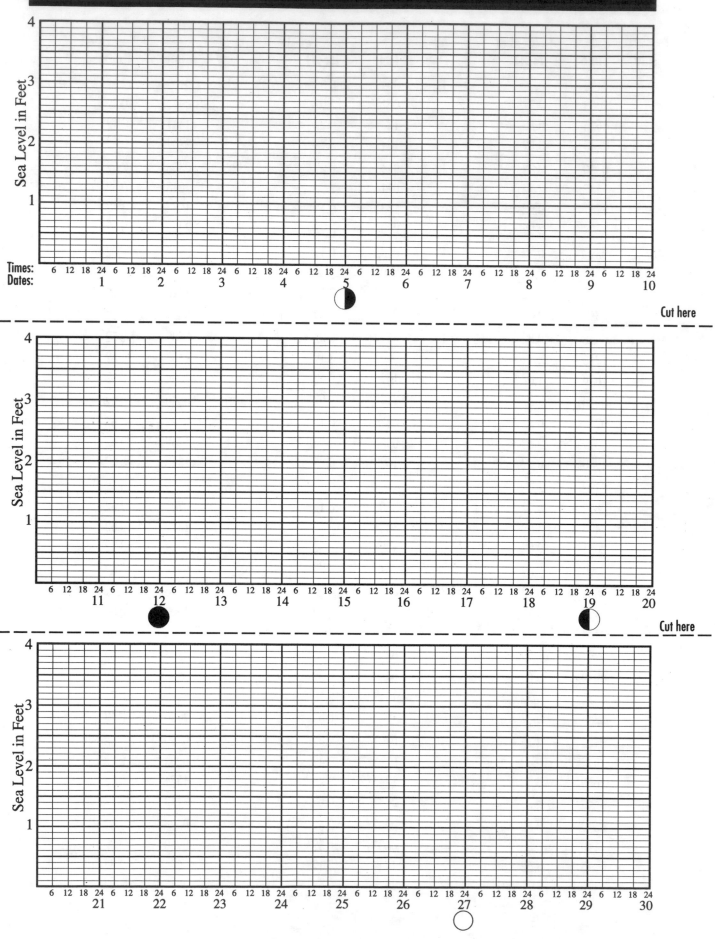

Cut here

Cut here

Plotting Tidal Curves

What is Happening?

Tides are experienced along the coastlines of the world as rhythmic fluctuations in sea level. They occur predictably at regular intervals. The pattern of tides, however, is different for every location. This variation from place to place in the pattern of tides is due to a number of factors, the most important being the nature of the land. By plotting the tidal data for a specific location, a picture of the pattern of tides for that place can be obtained and predictions for future tides can be made.

The reasons for tides are not discussed in this activity. They are discussed in Activity 15, *Tides Mobile* and Activity 16, *The Bulge on the Other Side of Earth*. Spring and neap tides are not discussed here either but are included in Activity 15. Questions 2 and 3 in the student section of this activity lead naturally into a discussion of spring and neap tides.

Materials

◊ pencil with eraser

◊ red pen or a bright color crayon

◊ scissors

◊ clear tape

◊ ruler

Important Points for Students to Understand

◊ Rhythmic fluctuations in sea level are called "tides."
◊ Tides occur at regular intervals. As a result, time and height of tides are predictable.
◊ The pattern of tides varies from location to location.
◊ Knowing tidal patterns is important for those who seek livelihoods or recreation from the oceans.

Time Management

Two class periods should be allowed for this activity. It will take students more than one class period to plot the data.

Preparation

Very little preparation is required for this activity. A ruler is not necessary. Any straight edge will work to help students connect the data points. Also note that feet are used not meters.

Suggestions for Further Study

The pattern of tides that the students will see in this activity is called "semi-diurnal," which means two high tides and two low tides each day. Have students investigate other tidal patterns, identifying factors that account for the different patterns.

The change in sea level could be used to generate electricity. This is actually being done in several places around the world. What are the factors that make a location on the coast suitable for such a project? What factors might limit the use of this technology?

Students often have the misconception that a "tidal wave" is caused by tides. "Tidal wave" is actually a misnomer. These waves have nothing to do with tides. The correct term is "seismic sea wave" or "tsunami." Students might investigate this phenomenon and learn why it is truly not a tidal wave.

Answers to Questions for Students

1. a. about twelve and a half hours

   b. about twelve and a half hours

2. a. June 12 and 13.

   b. June 12 and 13.

3. a. June 21–24.

   b. June 5.

4. a. Yes. It's low tide.

   b. High tide occurs at 2:03 p.m., so you have about two or three hours at the most.

5. 2:44 p.m.

From *Water*

There is a stream that flowed before the first beginning
Of bounding form that circumscribes
Protophyte and protozoon.
The passive permeable sea obeys,
Reflects, rises and falls as forces of moon and wind
Draw this way or that its weight of waves;
But mutable water holds no trace
Of crest or ripple or whirlpool; the wave breaks,
Scatters in a thousand instantaneous drops
That fall in sphere and ovoid, film-spun bubbles
Upheld in momentary equilibrium of strain and stress
In the ever-changing network woven between stars.

Kathleen Raine

Tides Mobile

Background

The Sun, Moon, and Earth are three extremely large objects separated by very great distances. (The Moon and Earth are 384,000 km apart and the Sun and Earth are 150 million km apart!) Despite the large distances between them, each object affects the others. Earth is kept in orbit around the Sun by the **gravitational forces** between them. The Moon is kept in orbit around Earth by the gravitational forces between them. These gravitational forces are mutual, meaning each object attracts *and* is attracted by the other.

Gravitational forces between the Moon and Earth create the tides. As you investigated in Activity 14, *Plotting Tidal Curves*, the coastlines of Earth experience tides as rhythmic fluctuations in sea level. Tides are caused primarily by two factors: the gravitational pull of the Moon and the inertia of water in the ocean on Earth. Gravity creates a bulge of water on the side of Earth facing the Moon which we will explore in this activity. (Inertia creates a bulge on the side of Earth facing away from the Moon and will be explored in Demonstration 16, *The Bulge on the Other Side of Earth*.)

The Sun also affects the tides. It creates two much smaller bulges, so small that, most of the time, they are not even noticed. As with the Moon, one is caused by gravity and always faces the Sun. The other is caused by inertia and is always on the side facing away from the Sun. However, four times a month, the Sun's effect on Earth's tides is noticeable. Twice a month, the Sun, Moon, and Earth align to produce very high and very low tides, called **spring tides**. Also twice a month, the three are aligned in a way that produces moderate tides, called **neap tides**.

In order to see how the Sun, Moon, and Earth interact to create tides, you will build a model of the three objects to illustrate the relative positions between them and how these positions change over the course of a month.

Objective

The objective of this activity is to construct a mobile that shows the relationship between the Sun, Moon, and Earth and to use this mobile to investigate how tides are created.

Materials

◊ scissors

◊ coat hanger

◊ string

◊ meter stick or dowel

◊ modeling clay

◊ tape

◊ yellow construction paper

◊ pencil

◊ paper clip

Vocabulary

Gravitational forces: The *mutual attraction between two objects*

Spring tides: Twice-a-month tides that are higher than usual.

Neap tides: Twice-a-month tides that result in a lower high tide and a higher low tide.

Procedure

Building the mobile. (See Figure 1.)

1. Begin by cutting a replica of the Sun out of construction paper. Tape this to the end of a piece of string and then tie the string to the middle of the meter stick (or dowel).

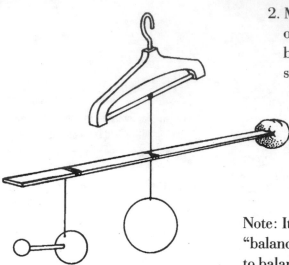

Figure 1

2. Make two balls of clay, one much smaller than the other. They represent the Moon and Earth. Attach a ball to each end of a pencil. (See Figure 3 for relative size and placement.)

3. Place a partially unfolded paper clip (see Figure 2) in the larger ball of clay as shown in Figure 3. Tie a piece of string to the paper clip after it has been inserted in the clay. Hold the string by the free end. If the pencil is not level, adjust the sizes of the clay balls until it is.

Note: It is important to realize that the Moon and Earth are "balanced" the same way as the two clay balls you are trying to balance. The situation is like a see-saw; a heavier person must sit closer to the balance point than a lighter person for the two to balance each other. Earth is so much "heavier" than the Moon (just like the clay models) that the balance point is located inside of Earth. Earth and the Moon rotate around this common point just like the clay models do.

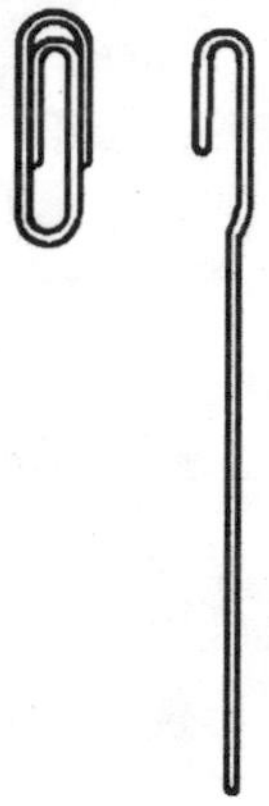

4. Tie the string to the meter stick at a point two pencil lengths from the string attached to the Sun. Adjust the level of the string so that the Moon and Earth are at the same level as the Sun.

5. Tie one end of another piece of string to the center of the meter stick. Tie the other end to the center of the horizontal piece of the coat hanger.

Figure 2

6. On the opposite end of the meter stick from the Moon and Earth, attach enough weight (clay, for example) so that when the mobile is held by the coat hanger, the meter stick is level as shown in Figure 1.

Experimenting with the mobile

7. Once the mobile is built, gently push the "Moon" so that the Earth-Moon system rotates. Then gently push one end of the meter stick in the same direction so that the whole mobile revolves around the Sun. This is a model of how the Sun, Moon, and Earth move relative to each other. (Obviously, though, the size and distance of your model are not to scale.)

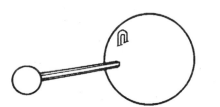

Figure 3

8. Figure 4 shows the part of the model with just the "Moon" and "Earth." With your pencil, shade in where the tidal bulges would be in this model.

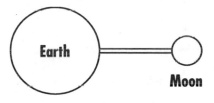

Figure 4

9. Stop the mobile from rotating. Arrange the Sun, Moon, and Earth so that the tides due to the moon and the tides due to the Sun add to each other making for a larger-than-usual tide. Is there another possible arrangement that would have the same or a very similar effect? Draw all the possible arrangements that would have this effect in the space below and explain why you have chosen them.

10. Arrange the Sun, Moon, and Earth so that the tides due to the moon and the tides due to the Sun "subtract" from each other. That is, in what arrangement would the tides due to the Sun and the Moon pull water away from each other? Is there another possible arrangement that would have the same or a very similar effect? Draw all the possible arrangements that would have this effect in the space below and explain why you have chosen them.

Tides Mobile

Materials

◊ scissors

◊ coat hanger

◊ string

◊ meter stick or dowel

◊ modeling clay

◊ tape

◊ yellow construction paper

◊ pencil

◊ paper clip

What is Happening?

The relationships between the Sun, Moon, and Earth have many effects on Earth. One of the most apparent is the formation of tides. Along the coastlines of the planet, tides are experienced as rhythmic fluctuations in sea level. While many factors act to create tides, two are most important: gravity and inertia.

Tides are due primarily to the relationship between the Moon and Earth. The Moon's gravity pulls the entire Earth toward it. Since water is a fluid and the rock and minerals that make up Earth are solids, the water is pulled further than the solids. This creates a bulge of water on the side of Earth facing the Moon. (The atmosphere also experiences "tides," but they are much more difficult to measure and do not affect us as much as the water-based tides.) The bulge on the other side of Earth is due to inertia.

One of the key points of this activity is to demonstrate to students that the Moon and Earth form a two-body system that rotates around a point. This point, however, is actually located within Earth at the center of mass of the two bodies. An effect of this rotation is that water tends to move, due to inertia, toward the side of Earth away from the Moon, creating another bulge. (This aspect of tidal formation is addressed in greater detail in Demonstration 16, *The Bulge on the Other Side of Earth*.)

These same factors—gravity and inertia—act between the Sun and Earth but to a lesser degree. Two much smaller bulges are created; one on the Sun-side of Earth and one on the other side. Normally, these bulges are not noticed within the tides. However, when the Sun, Moon, and Earth are positioned to form either a straight line or a right angle, the effect on the tides caused by the Sun is noticeable. When the three form a straight line—at new Moon and full Moon—the bulges due to the Sun enhance those due to the Moon, creating very high and very low tides. These are called spring tides. When the three form a right angle —at first and third quarter—the bulges due to the Sun detract from those due to the Moon, creating moderate high and low tides. These are called neap tides.

Important Points for Students to Understand

◊ The factors that are responsible for tides are due to the positional and gravitational relationship between the Sun, Moon, and Earth.

◊ The two factors primarily responsible for tides are gravity and inertia. These factors operate in the Earth-Moon system and the Earth-Sun system.

◊ Earth and the Moon form a two-body system which rotates about a point that is located within Earth.

◊ The tidal bulges due to the relationship between the Sun and Earth are not nearly as noticeable as those due to the relationship between the Moon and Earth.

◊ The tidal bulges due to the Sun are most noticeable when the Sun, Moon, and Earth are positioned to form either a straight line or a right angle.

Time Management

This activity can be conducted in one class period including time required for construction of the mobile. It is suggested that the mobiles be left up throughout instruction regarding tides so that the students can refer to them.

Preparation

Be sure that all materials are centrally located or already distributed to student groups. Any coat hanger will work for this activity. Some dry cleaners will give these away. If meter sticks are not available, dowels can be bought at most hardware stores. Modeling clay is sold at most craft stores. It may also be available from the art department.

Suggestions for Further Study

Tidal patterns vary around the world with regard to height and number of high and low tides per day. Several factors are responsible for this. The Bay of Fundy, for example has tides that are regularly 12 meters high while most locations have high tides of only a meter or so. Have students investigate why such variations occur.

Tides affect marine life in many ways. Sea turtles, for instance, only come ashore to lay their eggs during spring tides. Encourage students to find out why this is so.

The intensity of a hurricane's storm surge can be greatly increased if the hurricane makes landfall during a high tide. Have students investigate the relationship of tides and the impact of hurricanes on coastal communities.

Just as the oceans "bulge," so does the land, but this bulge is barely detectable. Have students explore methods of measuring Earth's bulges.

Answers to Questions in Procedure

8. Tides should be indicated as shown here.

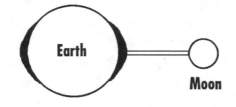

9. One possible arrangement is shown here.

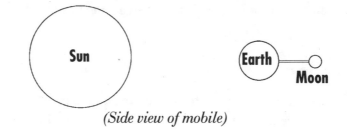

(Side view of mobile)

It would make the tides very high and very low because the bulges due to the Sun would add to the bulges due to the Moon. Another possible arrangement is shown below.

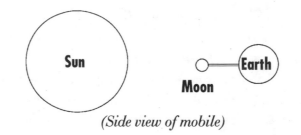

(Side view of mobile)

The tides would be higher and lower than usual for the same reason as above. Both of these arrangements create spring tides.

10. Both possible arrangements are shown below. In both cases the alignment would make moderate high and low tides because the bulges due to the Sun would detract from the bulges due to the Moon. These arrangements create neap tides. (NOTE: The following views are from above the mobile. Because a piece of construction paper is used to represent the "Sun" in this activity, it is represented by a line in the two drawings.)

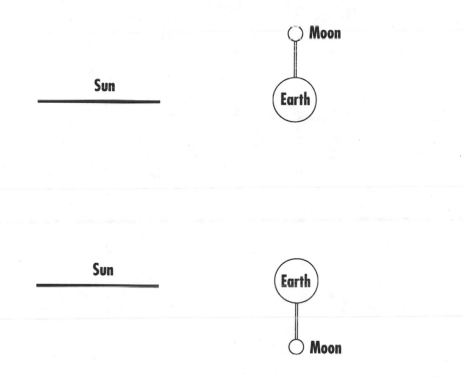

(Top views of mobile)

from *The Pleasure Boat*

And pictured beach of yellow sand,
And peaked rock, and hill,
Change the smooth sea to fairy land—
How lovely and how still!

Richard Henry Dana

The Bulge on the Other Side of Earth

What is Happening?

The bulge of water on the side of Earth that faces the Moon is easily explained. It is due to the gravitational attraction between the Moon and Earth, including the water on Earth. This attraction pulls water toward the Moon and creates a "bulge" on the surface of Earth.

The bulge on the other side of Earth is due to inertia. Inertia is the tendency of an object at rest to stay at rest and the tendency of a body in motion to continue its motion in a straight line. Pushing a stalled car is extremely difficult because such a massive object has a tremendous amount of inertia which must be overcome before it will move. Also because of inertia, the car is very difficult to stop once it has begun moving.

Inertia is the reason we wear seat belts in cars. When the car comes to a sudden stop, the people inside tend to keep moving. Objects moving in a circle (or any curved path) tend to keep moving, too. (See Figure 1.) But they tend to keep moving in a straight line, not a circle. (The law of inertia was devised to explain this phenomena.) Some force must be applied to keep the object from going in a straight line. Earth's gravity keeps the Moon moving in an almost circular orbit. Without the gravitational attraction between Earth and the Moon, the Moon would fly off into space. If a car goes around a curve, the passengers feel they are pushed against the door. Because of their inertia, they tend to keep moving in a straight line, and the door applies a force to make them follow the same curve as the car.

To see how inertia accounts for the bulge on the side of Earth away from the Moon, it is important to understand that Earth actually undergoes two types of rotation. First, it rotates on its own axis. Second, the Moon and Earth form a two-body system that rotates about a different axis. The axis is located at the center of gravity between the Moon and Earth. Because Earth is so much more massive that the Moon, the center of gravity is actually located within Earth.

Objective:

The objective of this activity is to demonstrate how the rotation of the Earth-Moon system accounts for the bulge of water on the side of Earth facing away from the Moon.

Materials

◊ Styrofoam ball (15 cm diameter)

◊ Styrofoam ball (6 cm diameter)

◊ string

◊ dowel (1 m long and 0.5 cm diameter)

◊ dowel (0.5 m long and 0.5 cm diameter)

◊ masking tape

◊ 2 weights (about 15 grams each)

◊ 473 ml drink bottle

◊ ruler

Vocabulary

Inertia: The tendency of matter to stay at rest or move uniformly along a straight line unless acted upon by an external force.
Center of gravity: The point in a body or system of bodies at which the entire weight seems to be concentrated.
Axis of rotation: A straight line about which an object or system of bodies rotates.

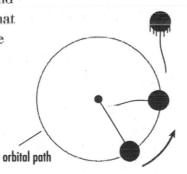

orbital path

Figure 1

Objects moving in a circle tend to move in a straight line. The ball in this diagram, when the string is cut, keeps moving in a straight line.

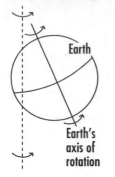

Center of gravity of Earth-Moon system and axis of rotation for Earth-Moon system

Earth

Earth's axis of rotation

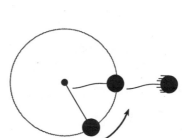

Moon

Figure 2

Figure 3a

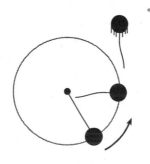

Figure 3b

The locations of the two axes and the difference between them is shown in Figure 2. An understanding of this second type of rotation is important in explaining the bulge on the other side of Earth.

This activity will demonstrate that the Moon and Earth form a two-body system that rotates about an axis located within Earth. The model illustrates the inertial tendency resulting from the rotation of the Earth-Moon system for objects (water among them) to move away from both sides of Earth—the side facing toward the Moon and the side facing away. Finally, the model demonstrates that the effect of things moving away from Earth is much greater on the side facing away from the Moon.

(NOTE: Many textbooks and other sources use the concept of "centrifugal force"—which is actually a misconception—to explain the effects of inertia described in this activity. According to this misconception, there is a force that acts on all objects which are in circular motion, and this force pushes or pulls the object out from the circle. There is no such force. The misconception arises from our own experience with circular motion. For example, when we go around a curve [part of a circle] in a car, we *feel* we are being pushed to one side of the car. This "feeling" has been incorrectly described as centrifugal force. The sensation is actually due to our bodies' inertial tendency to continue motion in a straight line rather than going around the curve.

The difference between the effects of inertia and the mythical centrifugal force are easily illustrated. Imagine a ball attached to a string that is swinging in a circle. If the string suddenly breaks, centrifugal force predicts what is illustrated in Figure 3a, which is incorrect. Inertia predicts a path for the ball as illustrated in figure 3b, which is correct. A little experience trying to hit a target with a sling will quickly prove that centrifugal force is a myth. Also see Reading 3, *The Tides: A Balance of Forces*.)

Important Points for Students to Understand

◊ Earth and the Moon form a two-body system that rotates on its own axis, independent of Earth's axis of rotation.
◊ The bulge on the side of Earth facing away from the Moon is due to inertia.

◊ Inertia contributes to both tidal bulges on Earth, but much more to the bulge facing away from the Moon.

Procedure

Part A

1. Assemble the apparatus as shown in Figure 4. The vertical dowel (the longer of the two) that runs from the bottle through the 15 cm ball should be inserted 2.5 cm from the surface of the ball, as shown in the diagram. The purpose of the masking tape is to keep the Styrofoam ball from sliding down the stick. Each string should be about 35 cm long. A simple way to attach the strings to the ball is to partially unfold a paper clip and stick the unfolded end into the styrofoam. (See Procedure 3 of Activity 15, *Tides Mobile*.) One end of the string can then be tied to the folded end and the other end to the weights. Heavy duty paper binder clips work well for the weights.

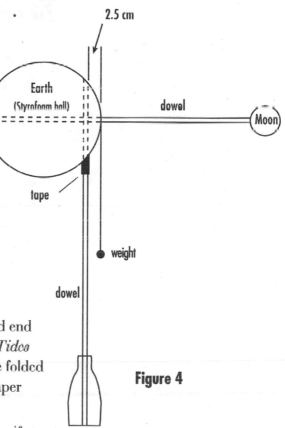

2.5 cm

Earth (Styrofoam ball)

dowel

Moon

tape

weight

dowel

Figure 4

2. Ask students to predict what will happen to the strings if you rotate the model. Holding the bottle with one hand, rotate the vertical dowel gently so that the whole apparatus spins. (Be sure to practice before doing it for the class.)

3. Ask the students to observe the weighted strings with regard to how close they come to being horizontal. Have them record their observations and draw a picture similar to Figure 3 showing the positions of the strings while the ball is spinning. If done properly, the string on the side opposite the "Moon" will come closer to horizontal than the other string.

Part B

4. Take the strings off the ball and re-attach them as shown in Figure 5.

5. After asking the students to make another prediction, spin the dowel again and have the students observe what happens to the strings. Have them record and draw their observations as before.

Part C

6. Remove the dowel and re-insert it in the center of the ball as shown in Figure 6.

7. Once the students have made predictions, spin the dowel and have the students record their observations as before.

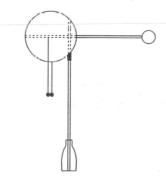

Figure 5

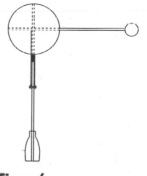

Figure 6

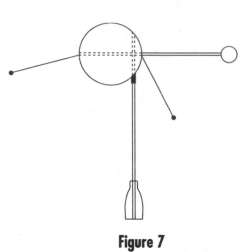

Figure 7

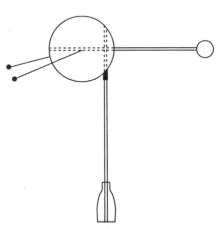

Figure 8

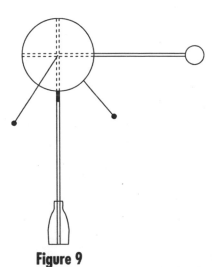

Figure 9

Discussion

The purpose of this activity is to show how the rotation of the Earth-Moon system accounts for the bulge of water on the side of Earth opposite the Moon. The water on the surface of Earth is rotating with Earth just as the strings rotated with the styrofoam ball. According to the law of inertia, the water tends to move in a straight line away from the surface of Earth rather than a circle, just as the strings moved away from Earth when the ball was spinning. Without Earth's gravity, the water in the ocean would fly off the surface and into space. If the strings had not been attached to the ball, they too would have flown off. The net effect of gravity and water's tendency to move in a straight line is the creation of a bulge of water on the side of Earth opposite the Moon.

Part A of this activity demonstrated several things. First, the strings moved away from "Earth" on both sides, but the string opposite the Moon moved farther away. Figure 7 shows what this might have looked like. This is because the axis of rotation is not located in the center of "Earth" but is shifted toward the "Moon." The same thing happens with the tides. A bulge due to inertia is created on both sides of Earth but it is much smaller on the side facing the Moon. This smaller bulge, however, adds to the one created by the gravity of the Moon making it roughly the same size as the bulge on the other side of Earth.

Part B of the activity, along with Part C, demonstrates another important aspect of tides. In Part B, the strings moved away from Earth, but they also moved away from the Moon, as shown in Figure 8. This effect is more noticeable when Part C is done. In Part C, the strings only move straight out from Earth, as shown in Figure 9. The difference is due to the location of the axis of rotation. In Part B, the axis was shifted toward the Moon. Everything on the side of the axis opposite the Moon moved away from the Moon. The same thing happens with the tides. All the water on Earth that is on the side of the axis opposite the Moon tends to move away from the Moon and toward the other side of Earth. This means that the inertial tendency of the majority of water on Earth is to move toward the side away from the Moon, and this explains why such a large bulge can be created there.

Because Earth has its own rotational axis independent of the rotation of the Earth-Moon system, this activity can be very confusing. A common misconception is that one of the tide-causing factors is Earth's own rotation. Many textbooks explain that tides are caused in part by Earth's rotation, but they fail to specify that it is Earth's rotation in the Earth-Moon system. The rotation of Earth on its own axis does not produce tides. Students and adults alike have a very difficult time distinguishing between these two types of rotation.

Preparation

It is possible to use two different sizes of sponge balls instead of the Styrofoam ones. The Styrofoam balls can be found in varying sizes at most craft stores. The apparatus should be set up before the demonstration. This gives the teacher an opportunity to practice spinning the dowel. Spinning it in such a way as to provide a smooth motion requires some practice.

Time Management

This demonstration can be done in less than one class period.

Suggestions for Further Study

Have two students stand facing each other, toe to toe. Then have them hold hands and, while keeping their backs and legs straight, lean backward. If students of different weights do this, one can see that one person will be leaning back farther than the other. That is, their center of gravity is not equidistant between them.

The water on Earth is not the only thing subject to tidal forces. The land is affected as well. However, since land is much more rigid than water, it is much more difficult to deform. In the same way, the Moon is subject to tidal forces. Have students investigate the objects in the solar system that are affected by tidal forces and learn what those effects are. In some cases, the effects are dramatic. (For example, research the Shoemaker-Levy comet impact—the comet broke up because of gravitational forces between it and Jupiter, the Sun, and Jupiter's moons.)

Encourage students to investigate on their own the difference between "centrifugal force" and inertia. They can do this by constructing a sling that will hold a tennis ball and trying to hit a target with it. It will not take many attempts to convince them that once released, the tennis ball does not fly out from the center of the circle but rather in a straight line tangent to the circle. (See Reading 3, *The Tides: A Balance of Forces*.)

The same factors that operate between Earth and the Moon to create tides also act between Earth and the Sun. Students can investigate how the factors are similar and different in the Earth-Sun system and what impact the factors have on tides on Earth.

Strandloper Stones

these rocks, heaved erect
in ancestral strength,
line the treacherous shores
that stretch this west-coast longitude.

the beach people
firmed their nomad tents
with whalebone struts,
fished first the seas
and drank from streams
deep within the dunes.

etched in indigenous light,
these beacons flashed
long before the Diaz myths,
heavy with salt crosses,
and skeleton ships
ribbed the Namib coast.

ringing now like battlements,
the circles claim their ritual space.

Dorian Haarhoff

Oily Spills

Background

Telephones, skis, antihistamines, ballpoint pens, music cassettes, toilet seats, and antifreeze. What do these items have in common? All are often made from oil. The United States and the rest of the industrialized world's economy is largely based on the products listed above and many more, all made from oil. This creates a huge demand for oil. To satisfy that desire, natural deposits of oil must be located and extracted. Currently much of the world's oil comes from areas such as Alaska and the Middle East, where these natural deposits of oil have been found and are extracted. The oil must then be transported over large distances in oil tankers or pipelines. When storing, handling, or transporting oil, there is always the chance of an accidental spill. Oil spills often cause serious damage to the physical and biological environments.

There are two properties of oil that contribute greatly to its damaging effects on the environment. The first is that unlike many substances, oil doesn't dissolve in water very well. The second is oil's tendency to float on the surface of water. Oil is less dense than water and less dense materials tend to rise above those that are more dense, like salad oil in oil and vinegar dressing (think back to the activities with the hydrometer earlier in this book). When spilled into water, large quantities of undissolved, floating oil forms slicks that threaten fish, water fowl, and other animals that live or breathe at the water's surface. Eventually much of the oil sinks to the bottom and forms "tar balls" that may wash ashore. Tar balls and slicks that wash up on land ruin shorelines, destroy plant and animal life, devastate wildlife habitats, and adversely affect human activities and industries.

There are several methods available for cleaning up oil spills in water or limiting their environmental damage. Floating booms are used to try and contain the oil. Detergents and other chemicals—known as **dispersants**—are used to act against the oil and break it up. The effectiveness of any method depends largely on the type of oil spilled, the size and location of the spill, and the prevailing weather conditions in the area. Other circumstances, unique to each spill, also affect the selection of a cleanup method or combination of methods.

In this activity you will simulate an oil spill in fresh water and experiment with various methods of oil spill cleanup.

Objective

The purpose of this activity is to explore the effectiveness of different methods for cleaning up oil spills.

Materials

You will work in groups of 4 or more for this activity. Each group will need:

◊ dish pan or plastic tub

◊ tap water

◊ vegetable oil or heavy olive oil (500 ml)

◊ cotton string (approximately 1 meter long)

◊ several drinking straws (cut in half)

◊ paper towels

◊ several small pieces of polystyrene foam (such as packing material)

◊ 100 ml of liquid detergent

◊ sand (125 g)

◊ diatomaceous earth (125 g)

◊ feather

Vocabulary

Dispersant: A chemical mixture that breaks up oil, keeping damaging slicks from forming and allowing the natural processes that degrade oil to occur.

Procedure

1. Fill the dish pan or sink half full with tap water.

2. Pour a small amount (slightly more than 100 ml) of oil into the water to simulate an oil spill. Note how a slick is created. (If necessary, replenish the water and oil in the pan or sink before trying each new method. Try to use nearly the same amount of oil each time.)

3. Construct a boom using the straws and string. (See Figure 1.) A boom is used to create a border around the spill in order to contain it. Try to contain or remove the oil using the boom. Record your comments about the effectiveness of this method in the Data Table.

4. Use a paper towel to gently blot or skim the oil from the surface of the water. Record your comments about the effectiveness of this method in the Data Table.

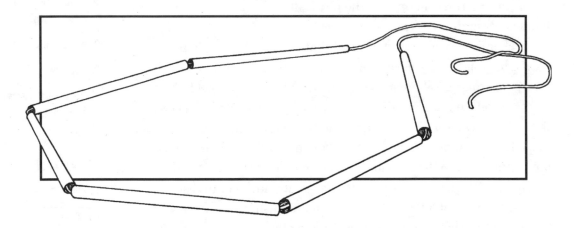

Figure 1

5. Place the pieces of polystyrene foam on the oil slick. After a minute or two, remove the polystyrene foam. Record what happened to the slick in the Data Table.

6. Add a drop of detergent to the surface of the oil near the center of the slick. Note what happens to the slick and record your observations in the Data Table.

7. Sprinkle some sand on the entire surface of the spill and note what happens. Record your comments in the Data Table.

8. Sprinkle some diatomaceous earth over the surface of the spill and observe what happens. Record your comments in the Data Table.

9. Devise one or more additional methods for cleaning up the oil. Have your teacher review your methods then try them out and record what happens in the Data Table.

10. In the Data Table, rank the effectiveness of each cleanup method by numbering the methods from 1 (most effective) to 5 (least effective)—if you devised additional methods, rank the methods 1 to 6 or more.

11. Dip the feather into the oil and note what happens to it. Record your observations in the Data Table.

12. Try rinsing the oil from the feather using tap water alone then tap water with soap. Record your observations in the Data Table.

13. When you have finished your investigation, clean the work area and discard used materials.

	DATA TABLE: OIL SPILL CLEANUP METHODS	
Cleanup Method	**Comments on Effectiveness**	**Method Rating**

Questions/Conclusions

1. Which cleanup method was the most effective in removing the oil spill?

2. What problems might be encountered when using this method to clean a real spill that has spread over a large area?

3. How might weather conditions affect the usefulness of this method?

4. Describe the problems you encountered using each of the other methods.

5. What factors (besides the size and location of the spill and the weather conditions) might affect the success of an oil spill cleanup operation?

6. Based on your observations of the feather after it was dipped into the oil, what do you think would happen to birds which were coated with oil as the result of a spill?

7. No attempt at controlling an oil spill is likely to be 100 percent effective. For example, in 1989, the *Exxon Valdez*—an oil tanker—ran aground on a reef in Prince William Sound off the coast of Alaska spilling more than 10 million gallons of oil into the water. It is estimated, for example, that less than one-half of the oil from the *Exxon Valdez* oil spill was recovered. The oil that remains in the environment will be slowly broken down by natural processes. It is possible to speed up the removal of oil from the environment using methods such as beach cleaning and addition of chemicals to the environment. These techniques are costly and time-consuming. Do you think the investment of time and money should be made to speed up the breakdown of oil in the environment? Why or why not?

8. What do you think are the causes of most oil spills? Can you think of some ways in which oil spills could be prevented?

Oily Spills

What is Happening?

The effectiveness of controlling and cleaning up oil spills depends largely on the type of oil spilled, the size and location of the spill, and the prevailing weather conditions in the area. In this activity, students will simulate an oil spill in fresh water and then evaluate various methods of oil spill cleanup. Remind students to draw on their knowledge from Activities 1, 2, 3, and 5 from this book which deal with the chemistry and structure of water to understand the relationship between oil and water.

Each of the cleanup methods students will use has a real-life counterpart. The straws and string used to make booms to contain the spill are very similar to the real life booms used to surround oil spills for containment. Paper towels model absorbent cloth which is used to soak up oil that has washed ashore. Polystyrene foam, sand, and diatomaceous earth all act to soak or attract the oil. Detergent acts as a dispersant help to disperse the oil so slicks do not form.

This activity also provides an opportunity for discussion on how oil that remains in the environment should be addressed. Natural processes will eventually degrade the oil, but the processes occur very slowly, on the order of years or decades. Physical cleaning can remove some of the oil from rocks and beaches. For instance, oil that reached shore from the *Exxon Valdez* oil spill was often cleaned by hand with absorbent cloth. Chemical dispersants can be used to speed up the degradation of oil. Microorganisms which consume oil are now used to combat some spills. Several of these cleanup techniques can be used on the same spill, but they are all costly and time-consuming. Debates arise as to whether such large investments into cleanup should be made and who is responsible for providing them.

Important Points for Students to Understand

◊ For the most part, oil and water do not mix because water is a polar molecule and oil is non-polar. Non-polar substances will not dissolve readily in water.

◊ Nearly all types of oil float on water because they are less dense than water.

◊ Many factors affect the selection of an oil cleanup method and the outcome of the cleanup operation.

Materials

Students will work in groups of 4 or more for this activity. Each group will need:

◊ dish pan or plastic tub

◊ tap water

◊ vegetable oil or heavy olive oil (~ 500 ml)

◊ cotton string (~ 1 meter)

◊ several drinking straws (cut in half)

◊ several small pieces of polystyrene foam (such as packing material)

◊ feather

◊ diatomaceous earth (~125 g)

◊ sand (~ 125 g)

◊ about 100 ml of liquid detergent

◊ paper towels

◊ Oil that remains in the environment after a spill may be left for natural processes to degrade or removed using costly cleanup methods.

◊ Oil spill cleanup is a complex, time-consuming, and messy activity.

Time Management

This activity can be completed in one class period.

Preparation

Prior to beginning the activity, collect and distribute the materials. If some of the materials listed are not available, the following items may be substituted: sponge or grass straw (both for absorbing the oil); aquarium net (for skimming). Alternatively, students may bring their own materials to try as oil removers.

Suggestions for Further Study

Simulate spill cleanup under varying weather conditions: add ice to the water to simulate an Arctic spill; modify the water temperature to illustrate differences between spills in temperate and tropical locations; use a portable hair dryer to illustrate complications of cleanup caused by wind. *(Caution: use care when using electricity near water. The use of a portable stand on which to support the hair dryer is recommended.)*

Use saltwater to simulate ocean water and compare cleanup methods for fresh and ocean waters.

Investigate actual oil spills around the world and the methods used to contain them. Also explore the economic effects on communities near the oil spills. (Examples include: the 1989 *Exxon Valdez* spill, a 1969 spill near Santa Barbara, CA, and the 1991 release of oil into the Persian Gulf during the Persian Gulf War.) In addition, investigate the impact oil spills have on sea life and coastal communities and debate who is responsible for cleaning up oil spills and who will pay the price of cleanup.

Answers to Questions for Students

1. Answers may vary. If different groups found different methods most effective, discuss possible reasons for the differences.

2. Answers will vary. For the boom method: it is difficult to create a continuous boom around a very large area. For absorption methods: covering the entire area may prove difficult. For dispersion methods: having a sufficient amount

of dispersant may be a problem and some dispersants may be harmful to fish and waterfowl. Also, dispersants are often applied by airplanes and can be deflected away from the spill by the wind as they are applied. All methods are affected by currents, obstructions (such as islands, shorelines, docks, buoys), and shallow water where cleanup vessels are unable to navigate.

3. All methods will be affected by extremely high or low temperatures, rough seas, high winds, storms, lightening, rain, hail, tides, and strong currents.

4. Answers will vary depending on observations.

5. Other factors affecting the success of cleanup are: how soon after the spill occurs that cleanup begins; the availability of the resources (people, money, and equipment) necessary to provide a continuous cleanup effort; the experience and training of the people involved; whether or not oil continues to spill after cleanup begins; how close to shore the spill occurs; the geography and geology of the area where the spill occurs (water temperature, an inlet versus open water, air temperature, the topography of the bottom).

6. Answers may vary. Birds whose feathers become coated with oil have difficulty flying, swimming, and gathering food. They may drown, be preyed upon, or starve to death if the oil is not removed. Also, their feathers lose much of their insulating properties and the birds may lose so much heat that they die of exposure to cold. They may also poison themselves as they try to clean their feathers and end up ingesting the oil.

7. Answers will vary. This question provides opportunity for discussion of the advantages and disadvantages of removing spilled oil from the environment.

8. Answers will vary. Some causes of oil spills are: tanker accidents (running aground, colliding with another tanker, split hull); pumping station accidents; leaks in storage tanks; and pipeline leaks. In coastal environments, small recreational boating spills are often a major source of spilled oil. Prevention of oil spills involves, among other things, special tanker construction to guard against leaks in the event of an accident; training of tanker, pump station, and pipeline personnel in the proper methods of handling oil-related equipment; and proper maintenance of oil-handling vessels and equipment.

On the Beach at Night Alone

On the beach at night alone,

As the old mother sways her to and fro singing her husky song,

As I watch the bright stars shining, I think a thought of the clef of the universes
of the future.

A vast similitude interlocks all,

All spheres, grown, ungrown, small, large, suns, moons, planets,

All distances of place however wide,

All distances of time, all inanimate forms,

All souls, all living bodies though they be ever so different, or in different
worlds,

All gaseous, watery, vegetable, mineral processes, the fishes, the brutes,

All nations, colors, barbarisms, civilizations, languages,

All identities that have existed or may exist on this globe, or any globe,

All lives and deaths, all of the past, present, future,

This vast similitude spans them, and always has spann'd,

And shall forever span them and compactly hold and enclose them.

Walt Whitman

Forever Trash

Background

Stories and legends speak of bodies and treasures buried at sea. Hundreds of thousands of boats and ships and the materials and supplies on them have sunk to watery graves since humans first sailed on the oceans. Even today, it is common practice for humans to throw their waste into the seas. In the past, much of that material would degrade and decompose in the ocean waters. The composition of waste, however, has changed from natural, **biodegradable** materials to synthetic materials that resist decay.

Use of products that do not decompose has increased dramatically over the past 20 years. As items made of plastic and plastic packaging have become more and more common, waste disposal has become increasingly difficult to control and regulate. Much of that waste ends up in oceans. The impact of synthetic debris on the marine environment has become painfully evident. Floating trash is visible in most bodies of water. Bits of plastic, polystyrene foam, aluminum beverage cans, and broken glass can be found on beaches throughout the world.

A series of laws passed by the U.S. government have outlawed dumping of medical waste, sewage sludge—a byproduct of sewage treatment—and plastics in the oceans and waters around the United States. And international treaties have tried to set the same standards for all countries, but ocean dumping and pollution remain a problem.

Plastics and other synthetic pollutants are made to last and do not break down readily. This "forever trash" floats in the surface waters, entangling and killing animals, or sinks to the ocean depths, altering the marine environment.

Procedure

1. Each group should select a designated number of items to "test."

2. Each item selected should first be subjected to a "sink or swim test." Record your predictions as to whether each item will float or sink in water in Table 1. Test your predictions by dropping the item to be tested in a beaker of water. Record your results in Table 2.

Objective

To observe the breakdown of various materials in water and in sand.

Materials

◊ small piece of paper (10 cm x 10 cm)

◊ 10-cm x 10-cm scraps of cloth (cotton, rayon, wool, polyester, nylon etc.)

◊ small sheets of aluminum foil, waxed paper, plastic wrap

◊ aluminum soda can pull tab

◊ plastic bag – sandwich size

◊ pieces of a plastic grocery bag

◊ plastic six pack holder

◊ plastic bottle cap

◊ hard candy in a plastic wrapper

◊ unwrapped hard candy

◊ rubber balloon

◊ polystyrene foam packing peanut

◊ sand containing organic matter

◊ shoe box or small individual containers

◊ beaker

Vocabulary

Biodegradable: Material that can be broken down into simpler substances (elements and compounds) by bacteria or other decomposers.

SCI LINKS.
THE WORLD'S A CLICK AWAY

Topic: ocean pollution
Go to: www.scilinks.org
Code: PES0155

3. Use the following instructions and test identical items for their ability to decay in water and in sand. Again, record your predictions as to the outcome of the tests in Table 1.

a) Place one of each item in a class aquarium filled with salt water. Leave the items in the water for one to two weeks.

b) Using individual containers or a "common grave" (shoe box filled with sand) bury one of each item in sand containing organic material. Leave the items in the sand for one to two weeks.

4. After the allotted time has expired, use a net to retrieve items from the water and a spoon to uncover the buried items. Observe and record the changes in each item in Table 2.

Questions/Conclusions

1. Which items floated?

2. List where floating trash might end up?

3. How can floating trash be harmful to marine life?

DATA TABLE 1: PREDICTIONS

	Item	Sink/Float	Description of Decay in water	in sand
#1				
#2				
#3				
#4				
#5				

DATA TABLE 2: RESULTS

	Item	Sink/Float	Description of Decay in water	in sand
#1				
#2				
#3				
#4				
#5				

4. Which items sank?

5. How might sunken trash be a hazard to marine life?

6. List the items that broke down or decayed most readily in the water decay test. Explain why these materials behaved as they did.

7. List the items that broke down or decayed most readily in the sand burial decay test. Explain why these materials behaved as they did.

8. Suggest several ways to limit the impact of pollution on our ocean.

9. How well did your predictions match the actual results you achieved? Explain the differences you observed.

Forever Trash

Materials

◊ small piece of paper (10 cm x 10 cm)

◊ 10-cm x 10-cm scraps of cloth (cotton, rayon, wool, polyester, nylon etc.)

◊ small sheets of aluminum foil, waxed paper, plastic wrap

◊ aluminum soda can pull tab

◊ plastic bag — sandwich size

◊ pieces of a plastic grocery bag (There are different kinds—some claim to degrade in light or landfills. Try to find an example of the different types.)

◊ plastic six pack holder

◊ plastic bottle cap

◊ hard candy in a plastic wrapper

◊ unwrapped hard candy

◊ rubber balloon

◊ polystyrene foam packing peanut

◊ sand containing organic matter

◊ shoe box or small individual containers

◊ beaker

◊ salt (for salt water)

What is Happening?

Trash and debris pose a serious threat to marine wildlife, navigation, and general water quality. Merchant ships, the world navies, commercial fishermen, recreational fishermen, boaters, and beachgoers all contribute to this problem. Human trash, in the form of food, plastics, beverage cans, polystyrene foam, cigarette butts, glass, etc., is an increasingly obvious characteristic of marine environments.

Scientists have identified plastic as the single most dangerous threat from humans facing many animals. Production of plastic in various forms has increased dramatically in the past 20 years. Plastic's durability makes its disposal extremely difficult. As a result, plastic has become an important marine pollutant. Plastics are often ingested by marine animals or serve as death traps—entangling, strangling, and suffocating the animals.

This activity explores various types of marine debris. Students will predict and discover which items float or sink and will observe the rates of decay of several common marine pollutants. Remind students that there are steps being taken to address the problem of marine pollution. Laws and treaties (discussed in Reading 5, *The Ocean: A Global View*) are good examples. Recycling programs for plastics, metal, paper, and other materials are also helping to reduce debris. Reiterate to students that they too can play a role in addressing the problem through thoughtful purchasing and understanding the science behind the problem and solutions.

Important Points for Students to Remember

◊ Rates of decomposition are different for different materials.
◊ Plastics are synthetic materials designed to last. They do not break down readily. "Degradable" plastics do not necessarily break down completely.
◊ Plastics, regardless of breakdown properties, remain a serious threat to marine wildlife.

Time Management

The set-up of this activity can be completed in one class period, with the same amount of time required for the wrap-up session.

At least one to two weeks is required between sessions—and it may actually take six to nine weeks for noticeable decomposition to occur.

Preparation

Obtain a large quantity of sand in which to bury waste materials. This sand should contain organic matter. If only commercial clean "sandbox" sand is available in your area, leave the sand outside in a bucket or large shallow pan for several weeks or add a small amount of organic material, such as pond water, to it. The amount of organic material in the sand will greatly influence the rate of decay in this experiment.

A large receptacle such as an aquarium may be used to model the decay of floating or sunken trash in the ocean. The addition of dissolved salts is suggested in modeling an ocean environment. (Note: This experiment may develop an unpleasant odor, depending on the presence of airborne particles and the composition of the waste materials being tested. A one-week testing period may be too long for some noses.)

Suggestions for Further Study

Some students should add an experimental control by testing material strength prior to decay. Pre-decay breakage, however, may be difficult to achieve with some materials..

Quantitative tests of decay may be made using equal-sized sheets of cloth, paper, plastic wrap, and aluminum foil. Simply bury each material in sand or water for an equal period of time. After the time has expired, retrieve the material and test its strength. To accomplish this, have two students pick up the material to be tested by all four corners; nothing should be supporting the test material. A third student should place a single penny in the center of the test material. Continue adding pennies one by one until the material breaks, tears, or disintegrates. Record the number of pennies necessary to "break" the material. Repeat this procedure for each material being tested. (NOTE: Materials suggested for quantitative studies include: different brands of paper towel; plastic wrap; aluminum foil; waxed paper; newspaper; notebook paper; and cotton, rayon, wool, or nylon cloth.)

Many states participate in an annual International Coastal Cleanup day, usually during the fall. Coordinated by the Center for Marine Conservation and other state and private agencies, the event focuses on cleaning up trash from along beaches and other water ways. Contact appropriate agencies in your state to organize or participate in the cleanup or similar activities. You may want to plan a cleanup near your school or in your local community.

Suggestions for Interdisciplinary Reading and Study

For years, the sea has held mystery and fascination for humans. Observe people who visit the ocean and you will see that they spend a great part of their time looking seaward—searching. Are they looking at incoming waves to check for swimming conditions, seeking out a speck on the horizon that will later materialize as a ship, or searching for treasures washed up by the waves? Or do people become spellbound by the soothing sight and sounds of ocean waves repeatedly washing upon the beach? While experiencing the tranquility of watching the sun rise or set across the ocean waves, do they become lost in the sheer mystery and vastness of the blue horizon? The sea has become the focus for expression in art, music, and literature.

To provide students with an opportunity to experience the beauty and value of the ocean and come to understand the importance of keeping the ocean and shorelines free of "forever trash," have students use sandpaper, crayons, and an iron to create a "sea print." Following are instructions for creating a print:

1. Use crayons to create a sea picture on a sheet of sandpaper, firmly pressing on the crayons to get a thick coat. After the scene is completed, fill in the background with an appropriate color.

2. Place the colored sandpaper face up on a thick stack of newspapers.

3. Place a sheet of manila paper over the print. Center the paper so that the print is in the middle.

4. Slowly press the print with a warm iron until you are able to see evidence of the crayons being transferred to the manila paper.

5. Trim and frame the print with a contrasting piece of construction paper.

Explore the economic and political impact of other types of environmental issues through role-play activities.

Answers to Questions for Students

1. Answers will vary depending on items selected. Some items float initially and then become saturated and sink.

2. Floating trash can wash up on beaches and become entangled in boat propellers, trapped among aquatic plants, or ingested by marine wildlife.

3. Plastics and other floating trash items are often ingested by marine animals. They have no nutritional value and interfere with the digestive process. Floating debris may also trap, strangle, or injure marine species.

4. Answers will vary depending on items selected. Some items float initially and then become saturated and sink. The size and the orientation of the material (for example, crumpled versus flat aluminum) may also effect whether a material "sinks or swims."

5. Sunken trash may be ingested by marine life, interfering with the digestive process. Sharp edges, rough surfaces and narrow openings may harm or injure wildlife.

6. Answers will vary depending on items selected and the length of the test period. Natural products decay much more readily than synthetics. Again, size and the orientation will effect the rate of decay.

7. Answers will vary depending on items selected and the length of the test period. Natural products decay much more readily than synthetics. Again, size and the orientation will effect the rate of decay.

8. Answers will vary. Some ways to limit the impact of pollution on our ocean include: use of recyclable and reusable materials, use of truly biodegradable materials, preprocessing waste before dumping, establishing and enforcing government restrictions on ocean dumping, and using alternative disposal methods (landfills, etc.).

9. Answers will vary.

Introduction

The following readings elaborate on the concepts presented in the preceding activities. While they were written especially for this volume with the teacher in mind, students should be encouraged to read them for interest and for additional study.

Water: The Sum of Its Parts

The oceans of the world are commonly thought of as distinct bodies in different regions of the globe. In reality, they are all part of one great world ocean. The concept of a world ocean is important to an understanding of world ecology and the effects that activity in one region have on environments and populations elsewhere. In keeping with this view, and to reinforce the notion of one world ocean, the term "ocean" as used in these readings generally refers to all of the world's oceans.

The beauty and mystery of the ocean have long fascinated children and adults alike. Studies of the ocean have revealed its usefulness, complexity, and importance to Earth's economic and ecological systems. The ocean supplies food, water, chemicals (such as iodine, bromine, and magnesium), fuels, recreational opportunities, and transportation for 90 percent of the world's international trade. In the United States, coastal regions make up only 15 percent of land area, but over 33 percent of the human population lives in coastal counties and 40 percent of manufacturing occurs there. The five largest cities in the world are located on coastlines. Many countries' communication and defense systems are largely ocean-based. These facts suggest a worldwide economic dependence on the ocean.

Perhaps less obvious than its economic significance are the effects the ocean has on Earth's ecosystem. The ocean occupies over 70 percent of Earth's surface and represents a rich and diverse biome. There is geologic evidence to suggest that life originated and developed in the ocean long before appearing on land. Microscopic organisms which thrive in the ocean supply more life-sustaining oxygen to Earth's atmosphere than any other source, including plants on land. The ocean's role in Earth's weather and climate systems is critical. It serves as a principal source of water for precipitation. It absorbs radiant energy from the sun, thereby moderating Earth's climate, especially in coastal regions. Ocean currents and wind distribute much of this absorbed heat throughout the globe, thereby affecting the climate in all regions of the world, even those remote from any coastline.

Henry Bryant Bigelow, the first director of the Woods Hole Oceanographic Institute, once said, "The most important thing about the ocean is that it is full of water." Seawater does indeed contain 97 percent pure water, and this statement calls attention to the crucial role the properties of water play in determining the properties of the ocean.

Water is a unique substance with properties much different from those of other compounds. It is the most common substance on Earth, yet it is the only substance that occurs naturally in all three states, or phases—solid, liquid, and gas. It is also unique in that its solid form is less dense than its liquid form—ice floats in water. Water has a high ability to store heat and dissolves more substances and in greater quantities than any other liquid. These properties have a tremendous impact on Earth and account for many of the life-sustaining attributes of the planet.

The water molecule is a fairly simple molecule made up of two elements—hydrogen and oxygen. Two hydrogen atoms (H) bond to one oxygen atom (O) creating a single water molecule (H_2O). To understand water's structure and why it exhibits certain characteristics, we must first examine an atom of each of its two elements. A hydrogen atom consists of one proton (in the nucleus) and one electron (moving around the nucleus). The larger oxygen atom has eight protons (and eight neutrons) in the nucleus, and eight electrons distributed within two shells, an inner shell with two electrons, and an outer shell with six electrons.

Outer electron shells that are completely full create a stable electronic configuration. Various elements, including hydrogen and oxygen, therefore interact with one another to form an arrangement that fills their outer shells. The outer electron shell of a hydrogen atom can hold up to two electrons and the outer electron shell of an oxygen atom can hold up to eight electrons.

Figure 1

A water molecule is stable because the outer shells of the two hydrogen and one oxygen atom are filled by the "sharing" of electrons.

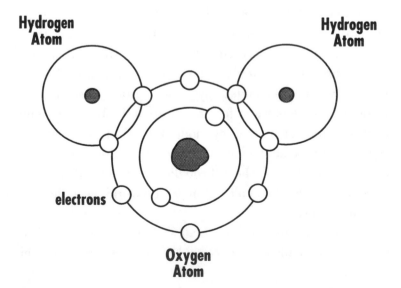

Hydrogen Atom

Hydrogen Atom

electrons

Oxygen Atom

The bonding that occurs between the hydrogen and oxygen atoms in a water molecule results from two hydrogen atoms each sharing an electron with a single oxygen atom so that each hydrogen has two electrons in its outer shell (one of its own and one "shared"), and the oxygen has eight electrons in its outer shell (six of its own and two "shared"). This "sharing" of electrons creates a very strong bond between the atoms called a **covalent bond.** The resultant molecule (H_2O) is stable because all of the outer electron shells of its constituent atoms are filled. See Figure 1.

The angle between the hydrogen atoms in a water molecule varies from 105° in the liquid phase to 109° in the solid phase, rather than being a 180° angle as might be expected. (The increase in the angle when water freezes results in an *increase* in volume and a consequent *decrease* in density—hence ice floats in water.) Also, the oxygen atom attracts the shared electrons more than the hydrogen atoms do. This results in an unequal distribution of electrons (and therefore negative charges) around the atoms. The shared electrons orbit closer to the oxygen nucleus than to the hydrogen nuclei. As a result, the water molecule is electrically polarized. The hydrogen end of the molecule is partially positively charged while the oxygen end carries a slight negative charge. See Figure 2.

The electrically polarized water molecule has properties something like a magnet—the positive end is attracted to the negative end of other molecules and vice-versa. This electrical attraction between water molecules leads to the development of cohesive forces called **hydrogen bonds**, which act between water molecules *and* between water molecules and other electrically charged particles. The degree of hydrogen bonding

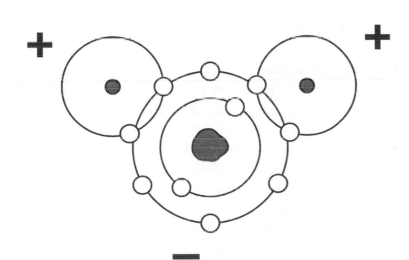

Figure 2

The oxygen atom holds the "shared" electrons more closely, resulting in a slight negative charge on the oxygen atom and slight positive charges on the hydrogen atoms. Because of this charge differential, water is termed a **polar molecule.**

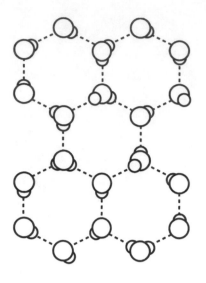

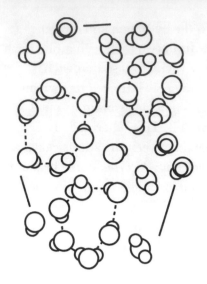

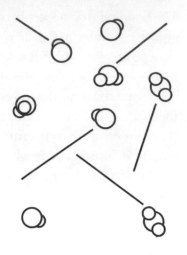

Solid
(many hydrogen bonds)

Liquid
(some hydrogen bonds)

Gas
(no hydrogen bonds)

Figure 3

that occurs in water varies with its physical state (solid, liquid, or gas).

In ice, water molecules are bound together by multiple hydrogen bonds into a nearly immobile crystalline structure of hexagonal shape. In liquid water, fewer hydrogen bonds exist and the water molecules slide past one another, allowing the water to "flow." In the vapor phase, no hydrogen bonds between water molecules exist and water is present as solitary molecules of gas. See Figure 3.

Consider that the temperature of water is a result of the motion its molecules exhibit. As more energy is introduced into a collection of molecules, the molecules move faster, hydrogen bonds are broken, and the temperature rises. Thus, as heat energy is added to ice crystals, they vibrate more and more until some of the hydrogen bonds are broken and the water molecules are able to move about in what is recognized as the liquid state. The molecules' ability to roll and slide past one another gives liquid water its fluid properties.

As heat energy is added to liquid water, the molecules move increasingly faster until, eventually, enough energy has been introduced to break the remaining hydrogen bonds between molecules. The molecules then move about randomly and have little attraction for one another; they have entered the gaseous state and exist as water vapor.

For the above process to be reversed, heat energy must be removed or released from a system containing water vapor. The speed of the molecules must be reduced to the point where hydrogen bonds form between some of the water molecules. In the atmosphere, for example, heat is released and molecular motion slows when water vapor condenses to form water droplets, as occurs in cloud formation.

As heat is removed from liquid water, increasingly more hydrogen bonds develop until the molecules are linked together in rigid hexagonal structures and ice is formed. Each of the processes that occur as water undergoes changes in state is shown in Figure 4.

Water is unique in that it is most dense in its liquid, rather than in its solid phase. For most other substances, the solid phase is most dense and therefore sinks in the liquid phase. Obviously this does not happen in water—ice floats. (Because water is such a common substance, many of us are more surprised that the solid phase of other substances does not float.) Ice forms at the top of a natural body of water, and stays at the top, resulting in a layer of insulation that prevents most lakes and ponds from freezing solid, even in the coldest weather. This layer of insulation also provides protection for the plant and animal life within the water beneath it. If solid water sank in liquid water, the world's

Figure 4

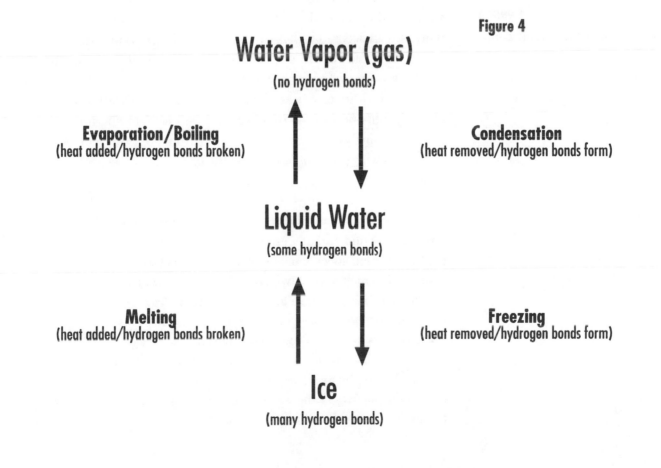

Water Vapor (gas)
(no hydrogen bonds)

Evaporation/Boiling
(heat added/hydrogen bonds broken)

Condensation
(heat removed/hydrogen bonds form)

Liquid Water
(some hydrogen bonds)

Melting
(heat added/hydrogen bonds broken)

Freezing
(heat removed/hydrogen bonds form)

Ice
(many hydrogen bonds)

lakes and ocean would tend to freeze solid, resulting in drastically different conditions on the planet.

The expansion of water upon freezing—ice occupies about 8 percent more space than the same amount of liquid water at 4 °C (pure water's temperature at maximum density)—plays a major role in the weathering and erosion of rocks. When the water in water-filled cracks freezes, the cracks enlarge and the rock is split further apart.

A common misconception surrounding the expansion of water upon freezing is that air fills the spaces within the hexagonal structure of the ice crystals and is responsible for making ice less dense than liquid water. In reality, the molecules in solid water (ice) arrange themselves in a crystalline structure where there are fewer molecules per unit volume than in liquid water. (Remember that the angle between the hydrogen atoms in a water molecule increases from 105° to 109° when water freezes.) Fewer molecules per unit volume results in lower density. See Figure 5.

Another important property of water is its relatively high specific heat. Specific heat is defined as the amount of heat required to raise the temperature of one gram of any substance by one degree centigrade. Upon the addition of heat energy to water, hydrogen bonds must be broken before a temperature increase can occur. The large number of hydrogen bonds in liquid water means that a relatively large amount of heat must be added to cause an increase in its temperature.

The high specific heat of liquid water, coupled with the large quantity of water in the ocean, allows ocean temperatures to change slowly and by small amounts throughout the year as compared with the temperatures of the continents. The ocean is generally warmer than the continents in winter and cooler than the continents in summer. Because it absorbs and retains so much heat, the ocean tends to act as a moderator of Earth's temperature and climate, especially in coastal regions. Also due in part to the high specific heat of water, ice sheets, icebergs, and ice floes tend to melt very slowly, a factor that also contributes to making climatic change a gradual process. (These formations are also slow to melt because of their mass, their locations in the polar regions, and their high reflectivity.)

In addition, water is often characterized as "the universal solvent." The phenomenal ability of liquid water to dissolve is due to the polarity of the water molecule. The charged regions of the molecule interfere with the attraction between oppositely

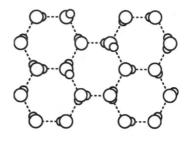

Figure 5

Water molecules arranged in a crystalline structure—ice.

charged ions within other compounds. If the attractions between the water molecules and the ions are stronger than the bonds between the ions, the substance will dissolve in water. For example, when NaCl (sodium chloride or "table salt") is exposed to water, the negative (oxygen) regions of the water molecules attract the positive (sodium) ions. At the same time, the positive (hydrogen) regions of the water molecules attract the negative (chloride) ions. The sodium and chloride ions are extracted from their crystal structure, the compound is broken down, and the salt dissolves. See Figure 6.

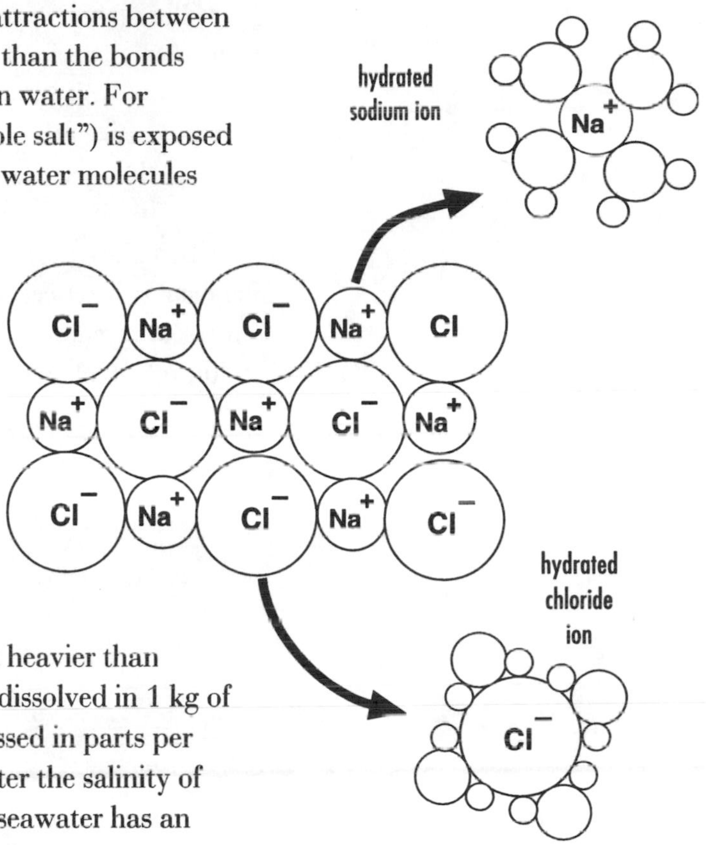

Figure 6

Water that contains dissolved substances has a greater weight per unit volume (density) than pure water. If 3 g of salt are dissolved in 100 ml of fresh water, the resultant mixture is 3 percent heavier than fresh water. The total salt content (in grams) dissolved in 1 kg of seawater is referred to as salinity and is expressed in parts per thousand (‰) by weight. Generally, the greater the salinity of seawater, the greater its density. Open ocean seawater has an average salinity of about 35 ‰ (or 3.5 percent).

The density of water is also affected by temperature. Liquid fresh water is most dense at 4°C, just before hexagonal ice crystals begin to form. As the temperature of liquid water increases from 4°C, its density decreases. All other conditions being equal, warmer water is less dense than cooler water.

Variation in seawater density occurs in both the open ocean and in coastal regions. In the open ocean, the variation results primarily from differences in water temperature; in estuaries and coastal zones, it results from differences in salt content. Because low-density water is more buoyant than water of higher density, low density water "floats" at or near the surface and water density increases gradually with depth. This **density gradient** in seawater bears important consequences that will be discussed in Reading 2, *The Ocean*.

Finally, one characteristic of water which is of interest, although it has few significant implications for oceanography, is its high surface tension. Surface tension (or cohesion) is the ability of a substance to adhere to itself. Water has the highest surface tension of all liquids. High surface tension allows some insects to remain on the surface of a body of water, even though their density is greater than that of water. All known marine-

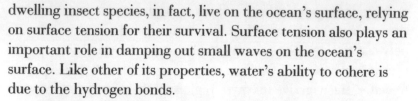

dwelling insect species, in fact, live on the ocean's surface, relying on surface tension for their survival. Surface tension also plays an important role in damping out small waves on the ocean's surface. Like other of its properties, water's ability to cohere is due to the hydrogen bonds.

The combined effects of the various characteristics described above account for the unique nature of water. This special substance, coupled with Earth's fortunate position within the solar system, was crucial for the emergence of life and plays a vital role in allowing Earth's species to thrive.

The Ocean

Just as water is unique among molecules, its existence in our ocean makes Earth unique among the planets. Our planet is the only body in the solar system that has a vast open ocean of liquid water. Mercury, Venus, and the Moon are completely dry. The only water found in the gas giants Jupiter, Saturn, Uranus, and Neptune is lost in mixture with greater quantities of hydrogen and helium. On the smaller worlds that lie beyond Earth, including Mars, Pluto, and the moons and other satellites of all of the planets, any water that has been detected is in the form of ice. It is Earth's vast ocean, in combination with its fortunate position in the solar system, that makes our planet life-bearing.

There are two main hypotheses presented to explain the origin of the world ocean, which now covers over 70 percent of Earth's surface. The first hypothesis proposes that water vapor was slowly released from molten material beneath Earth's surface by volcanic activity. As the concentration of water vapor in the atmosphere increased, some of the vapor condensed, clouds were formed, and rain began falling on the planet's surface. The second hypothesis suggests that most of Earth's water originated in comets, which were much more abundant in the first billion years of the planet's history than they are today. According to this hypothesis, as comets entered Earth's atmosphere, they instantly vaporized, adding moisture to the air that eventually fell to Earth as rain or snow.

For whichever reason, or combination of reasons, the rain began. As it continued, the low places on Earth's surface were filled with water and the ocean was formed. Rain has tended to mold Earth into a smooth sphere through the process of erosion ever since. Fortunately for us, tectonic forces within Earth keep raising up new land masses; otherwise, our planet would eventually be covered by a vast ocean, about 2400 meters deep, unbroken by continents.

Like any large environmental system, the ocean is usually subdivided into characteristic regions when studied scientifically. The most fundamental scheme of subdivisions is the one followed here, i.e., coastal ocean and open ocean, with the open ocean further divided into surface, transitional, and deep layers. These oceanic subdivisions differ from one another in many ways, but treating them separately should not obscure the fact that energy, matter, and organisms move from one subdivision to the other;

SCi LINKS.
THE WORLD'S A CLICK AWAY

Topic: oceans
Go to: www.scilinks.org
Code: PES0171

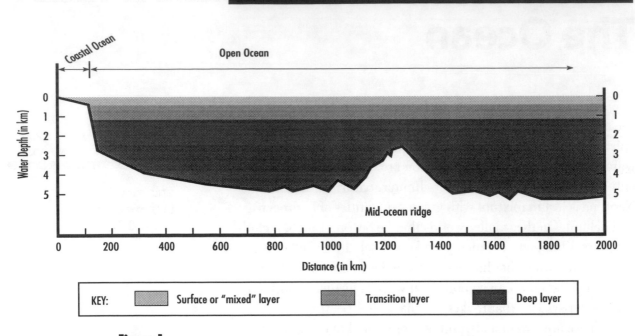

Open Ocean

Water Depth (in km)

Mid-ocean ridge

Distance (in km)

KEY: Surface or "mixed" layer   Transition layer   Deep layer

Figure 1
(Note: not to scale)

i.e., the ocean as a whole is an integrated environmental system even though scientists subdivide it for convenience of study. See Figure 1.

The coastal ocean refers to the 10 percent of ocean over the continental shelves (the undersea extensions of the continents which descend gradually and are not part of the deep ocean floor). It provides almost all of the world's seafood harvest, it is the source of all of the minerals and petroleum currently recovered from the ocean, and the coastline itself is home to a large and growing percentage of the human population. It is estimated that more than 60 percent of the population of the United States will live within 50 miles of the ocean or Great Lakes coastlines by the year 2010. The concentration of natural resources, human population, and economic activity make it no surprise that almost all conflicts and controversies regarding the use of the ocean originate from coastal ocean activities.

The single most characteristic feature of the coastal ocean environment is rapid change. Coastal winds are strong and variable, and they drive much of the coastal water circulation. Coastal winds also create waves that break on shore and move sediments, an action which can contribute to beach erosion. Coastal rainfall varies seasonally and within shorter time frames, and has a direct impact on the salt content of coastal waters. Variations in air temperature are generally reduced by the capacity of ocean water to store and release heat; but local and regional winds can overwhelm the moderating effect at relatively low speeds. Humidity, especially in summer, is generally high, and helps fuel potentially violent thunderstorms and hurricanes

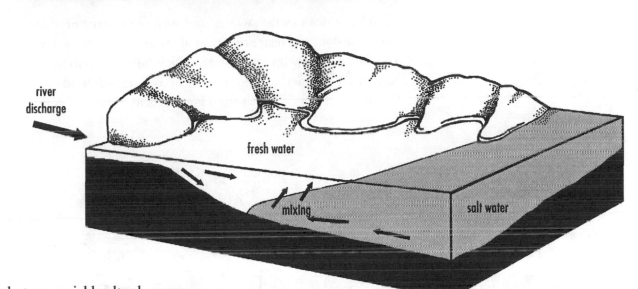

river discharge

fresh water

mixing

salt water

Figure 2
A typical salt-wedge estuary

that can quickly alter human activities and environmental conditions. Freshwater and sediment runoff from major rivers—especially after prolonged periods of rainfall—can also dramatically affect the coastal environment. The complexity of the coastal environment lends fascination to the phenomena that occur there, but may lead to oversimplification of ideas or the development of misconceptions among students.

An important feature of the coastal ocean is the **estuary**. An estuary is a semi-enclosed body of coastal water that exhibits measurably reduced salinity due to the introduction of fresh water from rivers, streams, and other sources of continental runoff. Most estuaries exhibit two-way movement of water. Less dense fresh (river) water flows seaward along the surface, while more dense, salty ocean water flows landward underneath it. Varied amounts of mixing of fresh and salt water occur, depending on factors such as wind speed, tide level, the depth and contour of the estuary bottom, and the relative inflow of river versus seawater. When mixing is minimal, a salt-wedge type of estuary develops, with a wedge of seawater undercutting an overlying wedge of fresh water. See Figure 2.

Estuaries and other coastal regions account for much of the biological productivity of the ocean. These areas are nutrient-rich from the accumulation of materials brought from the land via runoff; and they provide relatively sheltered habitats that are the spawning places and nurseries for many forms of marine life.

The remaining 90 percent of the ocean is open ocean—that is, it occurs seaward of the world's continental shelves. The open ocean is characterized by the surface layer, the transitional layer and the deep water zone. In the surface—or "mixed"—layer, turbulence due to wind, waves, and surface currents produces fairly constant physical properties of temperature and salinity. This layer is a few hundred meters thick (about 300 m in equatorial regions); its depth varies according to latitude and frequency of storms. The mixed layer acts as a thermal cap.

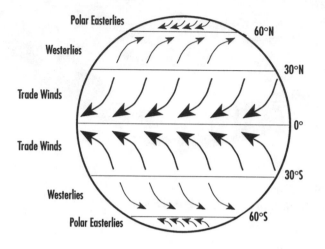

Global wind circulation patterns

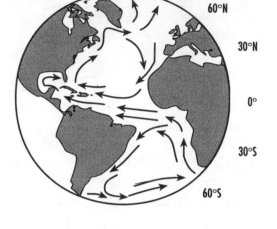

Surface ocean currents in the Atlantic Ocean

Figure 3

The relationship between global wind patterns and ocean surface currents.

Below it the temperature drops from whatever it is at the surface to the average temperature of deep water, which is a few degrees Celsius. This region of change is called the transition zone and is where we see the presence of a **thermocline** (rapid change in temperature), **halocline** (change in salinity), and **pycnocline** (change in density). The transition zone has a strong thermocline in equatorial regions where the difference between the surface and deep water temperatures is greatest. The thermocline weakens with increasing latitude and weakens seasonally in winter months in the mid-latitudes. The transition zone extends to a depth of about 1000 m. Beneath the transition zone, the physical properties of water are quite stable and uniform. This is the third layer, the deep water zone.

Among the winds that stir the currents are the trade winds and prevailing westerlies. These global winds blow in opposing directions and, in combination with Earth's rotation, cause a powerful system of rotating currents—called **gyres**—to develop. The water in the Gulf Stream region of the North Atlantic Gyre, for example, moves at speeds of up to 112 km/day. The correlation between global wind patterns and surface ocean currents can be seen in Figure 3.

The slowly circulating open ocean deep layer covers more than 75 percent of the area of the ocean basins, and more than half the total surface area of Earth. The deep ocean layer is the area of Earth about which the least is known. It has been determined, however, that the waters of the deep sea circulate in a pattern of convection currents which results from differences in the densities of masses of seawater. As discussed in Reading 1, *Water: The Sum of Its Parts*, the density of water increases with an increase in salt content (salinity) or a decrease in temperature; and more dense water tends to sink relative to less dense water. Therefore, seawater with high salinity sinks relative to seawater with lower salinity and cold water sinks relative to warmer water.

Much of the density variation in the deep ocean layer has it origin in the polar regions of Earth where seawater in the surface layer freezes and icebergs are formed. When water freezes to form ice sheets, which in turn sometimes break apart to form icebergs, the salt it contained remains dissolved and is concentrated in remaining unfrozen ocean water, increasing its salinity. (This is because the salt ions cannot easily fit into the crystalline structure of the ice.) In addition, the concentration of salts in the unfrozen water lowers its freezing point, allowing it to remain liquid below its normal freezing point of 0°C. This cold, salty water is more dense than the surrounding surface water and therefore tends to sink toward the seafloor and move away from the poles. Elsewhere, deep water is displaced upward toward the surface layer.

An example of a dense, deep-ocean water mass is the Antarctic Bottom Water (AABW). The densest water mass in the ocean, the AABW is created when icebergs form in the ocean around Antarctica, leaving behind extremely salty water with a temperature of less than 0°C. This water sinks to the seafloor as a coherent and identifiable mass. It creeps along the bottom of the ocean basin at speeds as slow as 30 m/day. Contrast this speed with the rapid—112,000 m/day—movement of the Gulf Stream surface current! The AABW has been traced as far north as Bermuda and France before converging and mixing with a much larger water mass from the North Atlantic. See Figure 4.

It is important to note that the convection currents that occur in the deep ocean layer are perpetuated by the freezing of surface water into icebergs near the poles. This leads to an important aspect of convection that should be pointed out to students: convection currents are not necessarily caused by the introduction of heat. Rather, they are caused by density variation within in a fluid. Variation in density can result from nonuniformity in temperature (due to uneven heating *or* cooling), salinity, or both. The deep ocean layer's system of density-driven flow (convection) is referred to as "thermohaline" circulation, since its is caused by differences in temperature (thermo) *and* salt content (haline) of oceanic water masses.

The concept of a three-layered ocean is important to a study of the ocean as a whole. It is also an essential feature of all mathematical models of Earth's climate, since the surface layer exchanges water and energy with the atmosphere, while the deeper layers do not. It is important to recognize that the layers are not totally distinct and that individual water molecules circulate throughout both. It may take hundreds of years, however, before cold, dense ocean water that sinks to the ocean floor near the poles returns to the surface layer again. Oceanographers and climatographers are therefore examining the feasibility of introducing atmospheric greenhouse gases, such as carbon dioxide, into the sinking waters, thereby removing them from exchange with the atmosphere for long periods of time.

Figure 4

Deep ocean currents in the Atlantic Ocean.

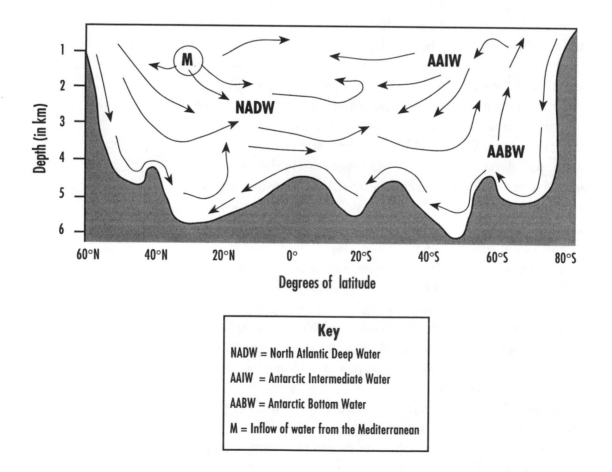

NATIONAL SCIENCE TEACHERS ASSOCIATION

The Tides: A Balance of Forces

Despite the fact that tides are a phenomenon familiar to most individuals, the driving forces behind the tides tend to be considered somewhat mysterious and confusing. Misconceptions about what causes tides abound and most people have only a vague awareness that the Moon is somehow involved. A global view of the tides can help explain the complex actions and interactions of the forces involved in tide formation.

The tides can be viewed as one global wave with two crests (or tidal bulges) and two troughs (or tidal troughs). Each crest

SCI LINKS.
THE WORLD'S A CLICK AWAY

Topic: tides
Go to: www.scilinks.org
Code: PESO177

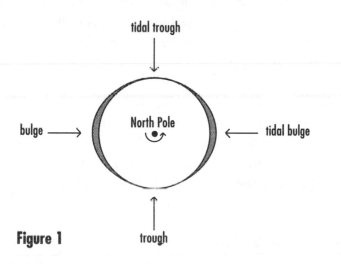

Figure 1

and trough runs the length of Earth from the North Pole to the South Pole. One bulge is always oriented toward the Moon, the other always away from the Moon. The global wave rotates slowly as Earth rotates much more rapidly underneath it. The different speeds of rotation makes the wave appear relatively stationery, and this view is therefore called the static model of tides. Figure 1 represents a simplified version of the static model.

To observers on Earth, the tidal bulges are experienced as high tides, the tidal troughs as low tides. The tides change as Earth rotates beneath the global wave. Because it takes 24 hours for Earth to make one complete rotation (360°) on its axis, it might be predicted that consecutive high or low tides would occur exactly 12 hours apart (the time it would take for Earth to complete one-half a rotation (180°) from one tidal bulge or trough to the other). In reality, the period between consecutive high or low tides is approximately 12 hours, 25 minutes. The period is 25 minutes longer than predicted because of the rotation of the global wave. The global wave rotates because the tidal bulge facing the Moon is drawn by the Moon's gravity as the Moon revolves around Earth. Because the wave rotates in the same direction as Earth but more slowly, it takes longer than 12 hours for a location on Earth to rotate from one point directly beneath a tidal bulge or trough to the next.

An analogy can help provide a clearer picture of this situation. Imagine that two people—a runner and a walker—are standing still at the same point on an oval, quarter-mile track. The runner can run around the track in one minute; the walker goes much more slowly. They both start around the track at the same instant and in the same direction—one running, the other walking. In one minute, the runner will be back at the starting point; but he will not have caught up with the walker, who has also moved around the track from the starting point. More time will be required for the runner to catch back up with (or "lap") the walker.

Similarly, on Earth, a location that experiences a high tide at 2:00 a.m. will rotate halfway around Earth in 12 hours. During that time, the global wave will have rotated to a much smaller degree in the same direction. Some extra time (25 minutes) will therefore be required for the location to move beneath the other tidal bulge and experience its second high tide. High tide will occur at 2:25 p.m. rather then 2:00 p.m. At 2:00 a.m. the next day, the location will have made one complete rotation, but the tidal bulge on that side of Earth will have continued to rotate away from its original position. The location will therefore not experience its next high tide until 2:50 a.m., 12 hours and 25 minutes after its previous high tide or 24 hours and 50 minutes after its initial high tide. See Figure 2.

Two concepts from physics are helpful in explaining tides. Newton's law of universal gravitation states that any two objects are attracted to each other by a force—gravity—that is directly proportional to the product of their masses and inversely propor-

Figure 2
Note: The dot along the edge of Earth in the figure represents the position of the observer.

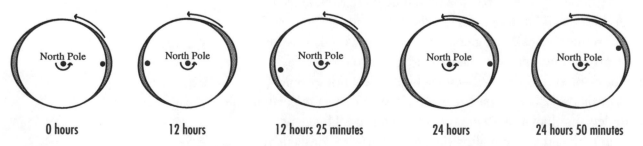

| 0 hours | 12 hours | 12 hours 25 minutes | 24 hours | 24 hours 50 minutes |

tional to the square of the distance between them. The greater the mass of an object, the greater its gravitational attraction; the greater the distance from an object, the smaller the effect of its gravity. Newton's law of inertia states that a body at rest tends to stay at rest, and a body in motion tends to stay in motion, in a straight line, unless acted upon by an outside force. These two phenomena—gravity and inertia—work together to create the tides.

According to Newton's law of inertia, the Moon, as it moves through space, has a tendency to continue its motion in a straight line and bypass Earth. However, according to Newton's law of universal gravitation, Earth and the Moon are attracted to one another due to gravity. Since Earth has much greater mass than the Moon, the effect of Earth's gravity is stronger. It acts in conjuction with the Moon's inertia to keep the Moon in orbit around Earth. Without the effect of Earth's gravity, the Moon would fly off in a straight line tangent to its orbit around Earth. See Figure 3.

The Moon also exerts a gravitational attraction on Earth. Land masses and the ocean are both affected by the Moon's gravity, but because water is more easily deformed than land, land masses are affected to a lesser degree and ocean water moves over the area of Earth facing the Moon. In this way, the Moon's gravity causes the tidal bulge which is oriented toward the Moon. See Figure 4.

It is commonly thought that the center of the Moon's orbit around Earth is in the center of Earth where Earth's center of mass and axis of rotation are located. In reality, Earth and the Moon together form a two-body system which rotates on an axis located at the center of mass of the system. The center of mass of the Earth/Moon system is distinctly different from Earth's center of mass. It is located within Earth, rather than at the center of Earth. This is because Earth is so much more massive than the Moon. The analogy of a seesaw can help make this point clearer.

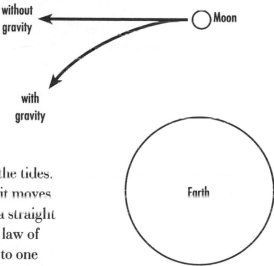

Figure 3

Figure 4

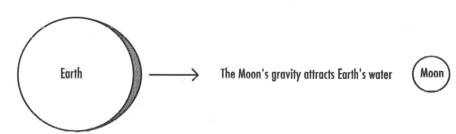

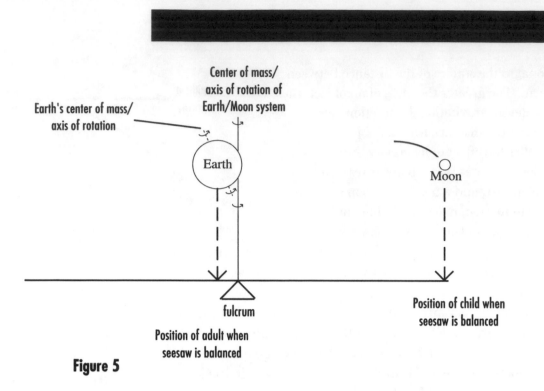

Center of mass/
axis of rotation of
Earth/Moon system

Earth's center of mass/
axis of rotation

Earth

Moon

fulcrum

Position of adult when
seesaw is balanced

Position of child when
seesaw is balanced

Figure 5

If an adult and a small child are on a seesaw, the adult must sit much closer to the fulcrum in order for the seesaw to be balanced. Similarly, if Earth and the Moon were placed on a seesaw, the fulcrum would have to be located within Earth for the two to balance. See Figure 5.

As the Earth-Moon system rotates about its axis, Earth, like the Moon, has an inertial tendency to continue in a straight line through space. Everything on the planet, including its ocean, is also subject to the effect of inertia. While Earth's gravitational attraction for the ocean keeps it from flying off the planet, it does move away from Earth somewhat because of its fluid nature. This is what creates the tidal bulge on the side of Earth opposite the Moon. The bulge is shown in an exaggerated way in Figure 6.

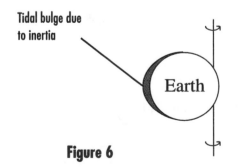

Tidal bulge due
to inertia

Earth

Moon

Figure 6

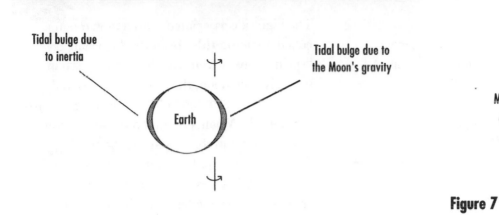

Tidal bulge due to inertia

Tidal bulge due to the Moon's gravity

Earth

Moon

Figure 7

The inertia of the Earth/Moon system contributes to the tidal bulge on the Moon side of Earth as well, but to a much smaller degree. Because the Moon side of Earth is very close to the center of mass of the Earth/Moon system, the force of inertia there is much less. Activity 15, *Tides Mobile* explores this difference.

In summary, the two tidal bulges on Earth result, for the most part, from two different factors. The bulge on the side of Earth facing the Moon is due primarily to the effect of the gravitational attraction of the Moon on Earth's water. The bulge on the opposite side of Earth is due primarily to the inertial tendency of water to travel in a straight line away from Earth as the Earth/Moon system rotates. See Figure 7.

The tidal bulges associated with the rotation of the Earth/Moon system should not be confused with the equatorial bulge that results from the rotation of Earth on its own axis. The equatorial bulge, although caused by inertia, is not a tide. It is a constant effect that occurs uniformly around the planet's equatorial regions. See Figure 8.

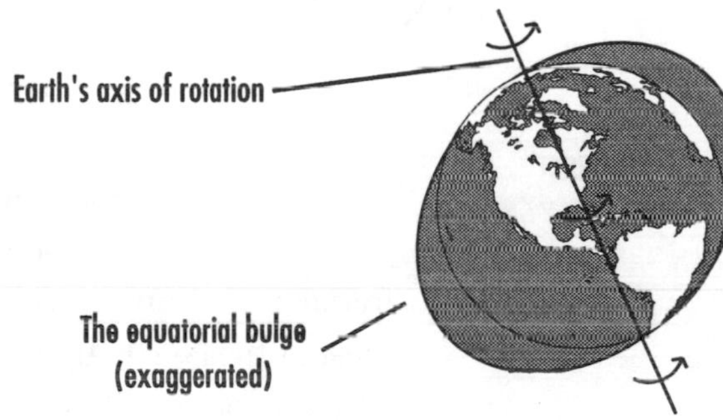

The ocean is also affected by the gravitational attraction of the Sun. While the effect is less than half that of the Moon's, it does becomes evident roughly four times a month—at new moon, first quarter, full moon, and third quarter. At new moon, the Sun's gravitational attraction works in conjunction with the Moon's to cause an exceptionally high tide.

Figure 8

Earth's axis of rotation

The equatorial bulge (exaggerated)

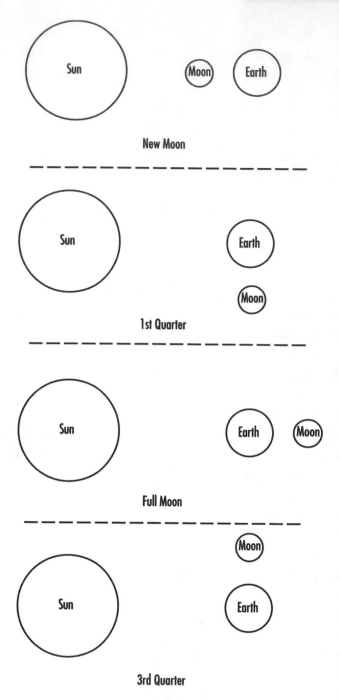

New Moon

1st Quarter

Full Moon

3rd Quarter

Figure 9

The high tide associated with a new moon is called a **spring tide,** because the ocean "springs" away from Earth even more than usual. At full moon, the Sun's gravitational attraction adds to the bulge on the side of Earth opposite the Moon, creating a second spring tide. At both first and third quarter, the Sun's gravitational attraction pulls water away from the normal tidal bulges and diminishes them. This creates the very moderate high and low tides referred to as **neap tides.** Spring and neap tides each occur twice every 29.5 days. See Figure 9.

Coastal locations of the world experience tides as rhythmic fluctuations in sea level. At high tide, the sea level rises; at low tide, the sea level falls. Not all locations experience the same pattern of tidal activity, however. Most locations on the eastern coast of North America, for example, experience diurnal tides—two high tides and two low tides every 24 hours and 50 minutes. Mixed tides—where there is a *high* high tide, a *low* high tide, a *high* low tide, and a *low* low tide every 24 hours and 50 minutes—occur along the west coast. Locations on the Gulf coast of Florida have a daily tide—one high tide and one low tide every 24 hours and 50 minutes.

Variations in tidal patterns result from the fact that the depth of the ocean is not uniform in all locations. Coastal ocean basins oscillate at frequencies either operating in conjunction with or opposing those of the basic tide producing forces. Just as the water in a bathtub sloshes back and forth at a frequency characteristic of its shape and bottom contour, the ocean water in coastal basins "sloshes" back and forth at a frequency unique to each basin. The coastal basin that affects most of the North America's eastern coastline has an oscillation period of about 24 hours and 50 minutes and therefore tends to reinforce tidal action. The reinforcing action is best seen in Canada's Bay of Fundy where oscillation and tidal periods reinforce each other to produce differences in sea level of up to 18 meters between high and low tides. Off the coast of California, the coastal basin oscillates in such a way as to reinforce every other tide, thereby creating the *high* high and *low* low tides. In the Gulf of Mexico, the basin oscillates once every 24 hours, thereby creating a single high tide and a single low tide each day.

A common misconception about tides that has even been perpetuated through some textbooks is that centrifugal force is in part responsible for their creation. The concept of centrifugal force is actually a myth. It is described as a force which pulls an object out away from the center of its circular orbit. The tidal and equatorial bulges are attributed to this nonexistent force through the assertion that water is being pulled away from Earth as Earth rotates. While the assertion that water tends to move away from Earth is true, the conclusion that centrifugal force causes the tendency is incorrect. If centrifugal force was at work, the water would tend to move along a radial path away from Earth; but the real tendency of water is to move along a path tangent to Earth. This tendency results from inertia, as previously described. Centrifugal force is not a real force, but a perceived effect. Some physics writings dispel the myth of centrifugal force and point out how it is misleading (see references below). The following is an excerpt from one such text (Ciancoli, 1980, page 66):

Figure 10

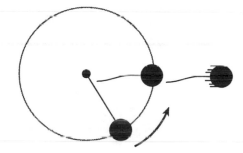

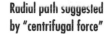

Radial path suggested
by "centrifugal force"

> "There is a common misconception that an object moving in a circle has an outward force acting on it, a so-called centrifugal ("center-fleeing") force. Consider, for example, a person swinging a ball on the end of a string around his head. If you have ever done this yourself, you know that you feel a force pulling outward on your hand. The misconception arises when this pull is interpreted as an outward "centrifugal" force pulling on the ball which is transmitted along the string to the hand. But this is not what is happening at all. To keep the ball moving in a circle, the person pulls inwardly on the ball. The ball, then, exerts an equal and opposite force on the hand [Newton's third law—for every action, there is an equal and opposite reaction], and *this* is the force your hand feels. The force *on the ball* is the one exerted *inwardly* on it by the person. For even more convincing evidence that a centrifugal force does not act on the ball, consider what happens when you let go of the string. If a centrifugal force were acting, the ball would fly outward. But it does not; it flies off tangentially, in the direction of the velocity it had at the moment it was released, since the inward force no longer acts." (See Figure 10.)

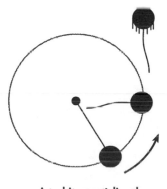

Actual (tangential) path

In teaching students about tides, it is very important that such misconceptions as centrifugal force be addressed. The difference between centrifugal force and inertia can be shown clearly through a simple demonstration. A student is given a ball attached to a string and asked to swing it about his or her head like a sling and to aim it at a target directly in front of him or her. The student soon discovers that centrifugal force is indeed a misconception. If the string is released when the ball is directly in front of the student (as centrifugal force suggests is necessary in order to hit the target), the ball will miss the target completely. With practice, the student will learn where the string must be released in order for the ball's tangential path to lead to the target.

In summary, while the tides are sometimes attributed to the effect of centrifugal force, this force is a misconception. The tides are actually generated by complex interactions of the forces of gravity and inertia inherent in the rotating Earth/Moon system. They can be viewed as a global wave under which Earth rotates, with crests experienced as high tides and troughs as low tides. The tides are reinforced or diminished by factors such as the natural oscillation of coastal basins and the gravitational effect of the Sun.

References

Giancoli, Douglas C. 1980. *Physics: Principles with applications*. Englewood Cliffs, NJ: Prentice-Hall.

Hewitt, Paul G. 1985. *Conceptual physics*. Boston: Little, Brown, and Company.

Taffel, Alexander. 1981. *Physics: Its methods and meanings*. Boston: Allyn and Bacon.

Waves

Ask anyone to name a characteristic feature of the ocean and, most likely, their answer will relate to waves. Waves figure significantly in our view of the ocean, because breaking waves are such a prominent feature along the beach. A careful review of wave characteristics, ocean wave formation, and types of ocean waves should address some of the natural curiosity, as well as some common misconceptions, that many students have about waves.

A **wave** is a disturbance in a medium that transmits energy from one place to another. This definition applies to light and sound waves as well as ocean waves. Terms commonly used to describe waves of all types include: **crest** – the highest point of a wave; **trough** – the lowest point of a wave;

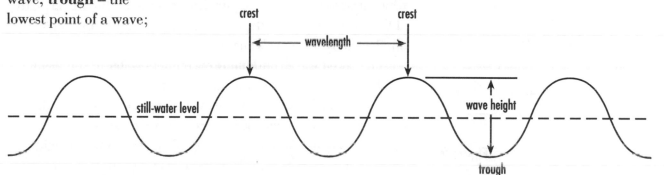

Figure 1

wave height – the vertical distance from crest to trough; **wavelength** – the horizontal distance between two consecutive crests (or two other similar points); **frequency** – the number of crests that pass a fixed point per unit of time; and finally, **speed** – the distance a wave travels per unit of time. See Figure 1.

There are a number of different types of ocean waves, the most familiar generally being those that break onto the beaches. Beach breakers are generated by the force of the wind. There are three characteristics of wind that influence the formation of a wind-generated wave: its duration (the length of time it has been blowing), its fetch (the distance over which the wind is blowing), and its average velocity (or speed). Fetch is related to the wavelength of the waves, while increases in the velocity and duration of the wind cause increases in the wavelength and height of the waves.

A common misconception among students is that the water that rushes ashore as a wave breaks is water that has traveled with the wave from many miles off shore. This, however, is not

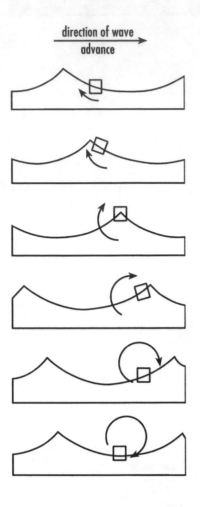

direction of wave
advance

Figure 2

Water particle
movement in a wave

the case. As a wave moves through water, it is only the wave energy or waveform that advances; the water particles (molecules) through which the wave passes move in a circular pattern and exhibit little or no overall forward motion. As the crest of a wave encounters a water particle, the particle moves to the peak of its path. As the crest passes by, the particle continues to the bottom of its circular path. See Figure 2.

Wind waves form because, as wind blows along the surface of a body of water, air pressure fluctuations and friction forces drag water into a series of crests and troughs. In the area directly affected by the wind, the waves—called sea waves—are irregular and travel in different directions. As sea waves move out from the area where they are generated, longer waves with higher velocities eventually overtake the shorter, slower waves. The long and short waves combine to form a regular pattern and become swell waves. The energy in a swell wave affects the water through which it passes to a depth equivalent to one-half the wavelength of the wave. Below this depth, water and the ocean bottom are unaffected by the wave's energy.

If a swell wave continually encounters water of sufficient depth, it can move across an entire ocean. As a swell wave moves into water roughly one-half of its wavelength or less in depth, however, the waveform encounters and begins to "feel" the bottom (seafloor). When a wave "feels" the bottom, its height increases and its velocity decreases. Because of the friction of water particles against the seafloor, the top of the wave moves ahead faster than the bottom of the wave. The wave eventually forms a curl of water that falls over forward, creating a breaking wave. See Figure 3.

If waves enter shallow water while moving at an angle to the beach or encounter irregular bottom features, the waves are refracted, meaning their direction of travel changes. The part of a wave that first encounters the bottom will be slowed down relative to the rest of the wave, causing the entire wave to turn (refract) toward the shallow water. In this way, waves tend to parallel shoreline contours. They converge on areas of the shore that protrude, and diverge from areas of the shore that curve landward. See Figure 4.

Water waves represent energy in motion. When a wave breaks, its energy is transferred to obstacles it encounters along the shoreline. The beach, which acts as a barrier to breaking waves, receives the brunt of their energy. Waves continually provide energy for the transport and redistribution of large amounts of sand through the shallow coastal waters. The energy

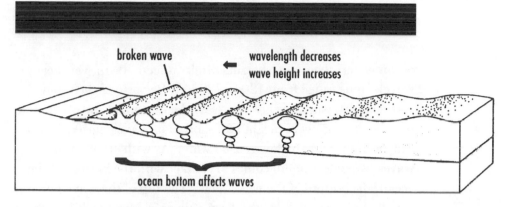

broken wave   ← wavelength decreases
                  wave height increases

ocean bottom affects waves

Figure 3

released from waves often leads to the development of local water currents. Waves that approach the beach at an angle, for example, can lead to the development of longshore currents (through which water is transported along a route parallel with the beach) and rip currents (through which water is transported seaward perpendicular to the beach).

Ordinary wind-generated waves cause continual, gradual changes in beach profiles over time through natural processes of sand deposition and erosion. Large, extremely powerful waves such as storm surges and tsunamis, however, can spread immediate devastation over large areas of the coastal environment. The center (or eye) of a hurricane is characterized by very low atmospheric pressure. Water is forced into this low-pressure area by the surrounding high-pressure air, resulting in a high water level in the center of the storm. This extra-high water is called a storm surge and can cause extensive flooding when a hurricane comes ashore.

Tsunamis, on the other hand, are great sea waves produced by the sudden displacement of a large volume of ocean water.

Figure 4

Refracted waves parallel the shape of the coastline.

headland

beach

Some causes of these sudden displacements are undersea earthquakes, landslides, and violent volcanic eruptions. As tsunamis travel through deep ocean water, their wave heights rarely exceed 30 cm, yet their wavelengths range from 120 to 720 km! Because of their long wavelengths, tsunamis always "feel" the bottom of the ocean, but they begin to grow in height and shorten in wavelength dramatically as they near shallow coastal water. As with other swell waves, when a tsunami comes in contact with the bottom of the ocean, the bottom of the wave slows. The top of the wave piles up a wall of water which brings devastating effects when it breaks on shore. Regions of the world at risk from tsunamis have developed a warning system to alert residents in the event of approaching danger. (The term tsunami is borrowed from the Japanese and translates to "harbor wave." The term "tidal wave" is often used interchangeably with tsunami; however, a tsunami is not shaped by the same forces as tides are. Therefore, use of the term "tidal wave" is discouraged.)

Water waves which are not ordinarily considered waves at all are the tides. An extensive review of tide formation may be found in Reading 3, *The Tides: A Balance of Forces*.

The Ocean: A Global View

The world ocean plays an integral role in the lives of all of Earth's inhabitants through its effect on global climate. As explained in Reading 1, *Water: The Sum of Its Parts*, the ocean's significance in climatic regulation is largely attributable to various characteristics water. The high specific heat of water means that a relatively large amount of heat energy must be absorbed or released by liquid water before a change in its temperature can occur. Because of the large volume of water in the ocean, ocean temperatures change slowly and vary less than the temperatures of land masses.

The ocean helps moderate temperatures over the globe through its effect on the conditions in masses of air that move across its surface. The air masses, when they move over land, moderate the temperature of land masses through precipitation and heat exchange. The moderating effect is most obvious in coastal regions, but regions far from any coastline also experience its influence. In fact, Earth's ocean causes the planet's overall temperature to be relatively consistent over time, especially when compared to that of other planets. On the other planets, and even on Earth's own moon, surface temperatures fluctuate much more widely during the course of a year, and in some cases, between day and night. Earth's relatively consistent temperature, maintained largely by its vast ocean, distinguishes it from its neighbors in the solar system and helps make it a habitable planet.

The ocean also affects global climatic conditions as a key component of the hydrologic cycle—the interactive cycling of water between the ocean, continents, and the atmosphere. In the hydrologic cycle, heat energy from the sun causes liquid water to evaporate and be introduced into the atmosphere as water vapor. Evaporation takes place from the surface of all bodies of water, as well as from soil and living organisms. Water vapor rises into the atmosphere, cools, and condenses to form the tiny water droplets that combine to make up clouds. When the amount of water in the atmosphere reaches a critical level, precipitation occurs and some of the evaporated water is returned to Earth. Because the ocean covers over 70 percent of Earth's surface, most precipitation falls there and mixes with seawater. Precipitation that falls onto land either evaporates directly back into the atmosphere, is absorbed by the soil, or flows into

Topic: water cycle
Go to: www.scilinks.org
Code: PESO189

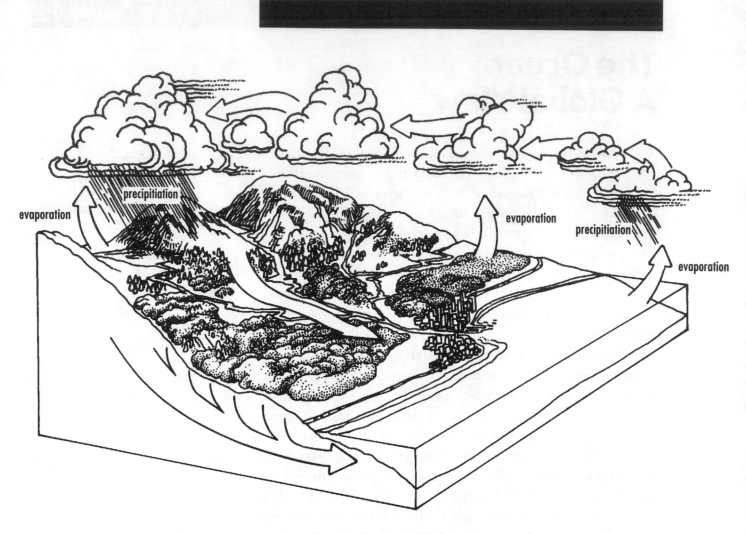

Figure 1

Earth's hydrologic cycle

streams. Small streams combine to form larger streams and rivers, which eventually flow into a lake or the ocean. All along the way, water is reintroduced into the atmosphere through evaporation, and the cycle continues. See Figure 1.

The salinity of the ocean, important for the survival of its inhabitants, is maintained through the hydrologic cycle. When rainwater falls on land, it erodes and dissolves salts and minerals from rocks and soil. These leached salts are carried by runoff through streams and rivers, and many eventually reach the ocean. It is estimated that some four billion tons of dissolved salts are carried into the ocean each year. At the same time, water that evaporates from the ocean leaves it salts behind, increasing the salt concentration of the remaining seawater. The ongoing introduction of salts into the ocean through these processes is offset by other events. Fresh water continually enters the ocean as precipitation and salts are deposited on the ocean floor at about the same rate as they are introduced into the ocean by rivers and streams. The balance between these opposing processes has caused the salinity of ocean water to remain essentially the same through recent geologic history.

Within the last few decades, scientists have developed a
fairly clear understanding of the role the ocean and the hydrologic cycle play in weather formation. One interesting result has
been the explanation of the effects of El Niño. El Niño is an
unusually warm ocean current that develops in the western
Pacific Ocean, usually around Christmas time (hence the name,
which means The Child). The warm surface water temperatures
associated with El Niño lead to alterations in the movement of air
masses and the development of irregular ocean currents. Changes
in the normal formation of weather conditions result, and the
entire world is eventually affected. In the mid-1980's, El Niño
was responsible for weather extremes from severe drought
conditions in Australia to heavy rains and flooding in parts of
South America.

The effects of El Niño provide evidence of the integral role
the ocean plays in the lives of human beings. As marine explorer
Jacques-Yves Cousteau stated in 1980, "The very survival of the
human species depends upon the maintenance of an ocean, clean
and alive, spreading all around the world. The ocean is our
planet's life belt." The importance of the ocean is often over-
looked, however, even as people around the world alter their
activities in response to daily weather conditions regulated in
large part by the ocean.

Similarly, the important impact that humans have on the
ocean is generally underacknowledged. The pollution that results
from certain human activities has the potential to permanently
alter the delicate ecological balance in the ocean and thereby
undermine its capacity to sustain life. For this reason,
Jacques-Yves Cousteau and many other noted scientists
advocate protection of the world ocean from overuse and from
various forms of pollution. A 1971 United Nations report defined
marine pollution as "the introduction by man, directly or indi-
rectly, of substances or energy into the marine environment,
resulting in such deleterious effects as harm to living resources;
hazards to human health; hindrance to marine activities, including fishing; impairment of quality for use of seawater; and
reduction of amenities."

One familiar ocean contaminant is oil. Oil seeps, in which
crude oil exudes from the ocean bottom, are common occurrences
in some areas. Natural degradation processes occur after an oil
seep that, given time, generally bring about recovery of the
affected area. Ocean pollution, on the other hand, arises from
human activity that introduces an "unnatural" amount of oil,

thereby overwhelming the capacity of the environment's degradation and recovery mechanisms. Oil is introduced into ocean water primarily during transportation activities. While in transport, oil may be released by "normal" leakage, accidents, or improper equipment cleaning methods.

Immediate damage to marine life results from an oil spill and occurs at the interface between the ocean and the atmosphere, where the oil forms a layer or "slick." As the *Exxon Valdez* accident of 1989 clearly demonstrated, seabirds, sea otters, and other aquatic animals that must surface for air are extremely vulnerable to the effects of spilled oil. The oil adheres to their feathers, skins, and coats and severely limits their mobility. They often become easy prey for predators or starve because they can no longer hunt successfully. Those that survive and are rescued must be individually cleaned and treated.

Compounding the threat of the oil slick itself is the fact that the hydrocarbon components of oil are extremely toxic to most forms of marine life. As oil that remains in the marine environment breaks down into its various components, some of the more volatile compounds evaporate into the atmosphere and others decay relatively quickly through natural processes. Other compounds, however, dissolve into the water or settle to the ocean bottom where they can enter the food chain if ingested by organisms. Through the food chain, these toxins are concentrated and their destructive effects are multiplied over time. Humans, at the top of the food chain, may be affected through this route.

Several techniques are used to contain and recover spilled oil. One technique involves specially equipped boats that skim the oil from the ocean's surface; floating booms can be placed around an oil slick in an attempt to contain it; chemical dispersants are sometimes sprayed over an oil slick to break down the oil and cause it to degrade faster; and, at times, the oil is even ignited. A new technique developed involves oil-eating bacteria. In many cases, however, efforts to clean up an oil spill meet with only limited success, and some clean-up techniques can pose greater threats to the environment than the oil spill itself. Ignition of an oil spill, for example, causes air pollution; chemical dispersants cause clumps of toxic material to sink to the ocean bottom where they are likely to be consumed by oysters, clams, and crabs.

Spilled oil that is not contained before it reaches shorelines can spoil beaches and thereby threaten the habitat of shorebirds, sand crabs, and other beach-dwelling creatures. Where possible,

SCI LINKS.
THE WORLD'S A CLICK AWAY

Topic: ocean pollution
Go to: www.scilinks.org
Code: PES0192

pools of oil are removed through the use of absorbents such as hay; rocks and other objects are scrubbed by hand.

Ocean dumping (the disposal of waste products at sea) is another major form of ocean pollution. The shallow waters off the coasts of many countries are often used as dumping grounds for various wastes. The choice to dump garbage at sea, as opposed to in a landfill, is usually based on economic reasons, as land is generally deemed more valuable than the ocean. The United States alone dumps over 45 million tons of waste into the ocean each year, and much of the material may be detrimental to the environment.

Among the many pollutants disposed of through ocean dumping is a wide variety of plastics. In 1988, almost 50 billion pounds of plastic were produced, and the amount is increasing. During the 1988 National Coastal Cleanup, 62 percent of the trash collected was plastic. In one survey from 1985, plastic made up 86 percent of the trash observed floating in the North Pacific Ocean. This points to one of the problems inherent in dumping plastics at sea—most of it floats. Another problem is its durability. Plastics are relatively immune to the natural process of decomposition, meaning that plastics put into the ocean can float around and cause problems for many years. Animals of all kinds die as a result of being tangled in plastic debris, and many animals mistake plastic for food and die as a result of ingesting it.

Ocean dumping also impacts humans in a variety of detrimental ways. Floating debris can damage recreational and fishing vessels. Insurance companies estimate that over $50 million has been awarded for boat repairs that resulted from damage incurred by debris. Refuse that washes ashore pollutes beaches and deters tourists, thereby reducing tourism revenues. The state of New York lost an estimated $1 billion in tourism revenue when medical waste and other rubbish washed up on beaches there. Incredible amounts of Federal and state revenue are spent to keep beaches clean and free of ocean debris. In 1988, for example, $1,275,354 were spent to maintain a three-mile stretch of beach in Santa Monica, California.

Legislation is aimed at restricting the ocean dumping that takes place in all navigable waters of the United States. Annex V of the Marpol Treaty designates how far out in the ocean certain materials must be dumped, and makes it illegal to dump any plastic anywhere in U.S. waters. While this type of legislation will undoubtedly have a positive effect, it represents only a small step toward alleviating the problems associated with ocean dumping. There are several other things being tried to improve the situation, the most promising being the reduction of plastics use and recycling. Improvement may also come from the development and use of naturally degradable plastics.

The ocean can be viewed as an integral component of the global environment. It affects temperatures and climatic conditions worldwide and is largely responsible for making Earth a habitable planet. The ecologically significant characteristics of the ocean stem from the delicate balance of its many natural systems. Human activities that disrupt this balance threaten to alter the environmental conditions upon which we depend for survival. Therefore, knowledge about the ocean and informed decision-making regarding its use are essential for the maintenance of a life-sustaining planet.

Master Materials List

Originally designed as a program in leadership development, the purpose of Project Earth Science was to prepare middle school teachers to lead workshops on topics in Earth science. These workshops were designed to help teachers convey the content of Earth science through the use of hands-on activities.

We suggest organizing *Project Earth Science: Physical Oceanography* activities around three key concept areas, explained further in the Introduction on pages 9–12. To assist workshop leaders, this Master Materials List includes all of the equipment necessary for using the activities under each concept area.

Concept Area I

water
several cups of ice
several cups of unflavored gelatin cubes—clear and colored
vegetable shortening
granular sugar
table salt (NaCl)
mineral oil (or baby oil)
isopropyl alcohol (70%)
Epsom salts (Mg_2SO_4)
baking soda ($NaHCO_3$)
sand
food coloring
scissors
glue
blank paper
black construction paper
crayons or markers (red and blue)
red wax marking pencil
string
ruler or meter stick
paper towels
three 250 ml beakers
240-ml (8 oz.) polystyrene foam cups
graduated cylinder
5 test tubes with rubber stoppers
test tube rack
5 containers varying in shape and size of opening
small beakers or clear plastic cups
teaspoon
small spoon (1/8 teaspoon will work)
spoon or tongs
hot plate with wire gauze screen
thermometer
balance
ring stand
utility clamps

Activities:

A Pile of Water
A Sticky Molecule
Over and Under—Why
    Water's Weird
How Water Holds Heat
Water—The Universal Solvent

lamp with reflector and 200-watt bulb
clock or watch with a second hand
eyedroppers
several rolls of pennies
other assorted coins (quarters, nickels, etc.)

Activities:

Won't You BB My Hydrometer?
Ocean Layers
The Myth of Davy Jones's Locker
Estuaries—Where Rivers Meet the Sea
Current Events in the Ocean
Body Waves
Waves and Wind in a Box
Tanks A Lot—Activities for
   a Wave Tank
Plotting Tidal Curves
Tides Mobile
The Bulge on the Other Side of Earth

Concept Area II

water
ice
food coloring (three colors: blue, red, green)
pickling salt
hard-boiled egg
2 kg of sand (optional)
masking tape
packing or duct tape
clear tape
modeling clay
slice of clay 3 cm thick
scissors
string
sheets of white paper
yellow construction paper
fine-tip permanent marking pen
red wax marking pencil
white chalk
colored pencils
black permanent marker
pencil with eraser
red pen or a bright color crayon
paper clip
plastic drinking straws with a flexible elbow
one or two large plastic trash bags (preferably white)
two cardboard boxes (75 cm x 28 cm x 5 cm or comparable size)
metric ruler
500 ml beaker
100 ml and 500 ml graduated cylinders
250 ml clear plastic cups
600 ml beakers (or jars that will hold more than 500 ml of liquid)
473 ml drink bottle
5 ml metric measuring spoon
eight small screw top vials (or 8 test tubes and stoppers)
large jar or beaker
three buckets (4-liter or 1-gallon size)
plastic cylinder with end caps (45 cm [18"] tall by 4 cm [1.5"] in diam.)
   for student setup
plastic cylinder with end caps (180 cm [6'] tall by 6 cm [2.75"] in diam.)
   for demonstration setup
soup spoon (large enough to fit egg on)
spoon
balance
plastic medicine dropper
three medicine droppers or plastic pipettes
clear plastic straw (about 10 cm long)
ring stand with clamp to hold cylinder

hot plate and pan or coffee heater
waste containers for used solutions
towels or rags for cleanup
cafeteria tray
large funnel
1 to 2 m length of rubber tubing with U-shaped glass tubing in one end
stopwatch or watch with second hand
meter stick or dowel
BBs
clear Pyrex™ glass loaf pans
baking pan, 30 cm x 45 cm x 3 cm deep (12" x 18" x 1.5"), painted black inside
400 ml of rheoscopic fluid
two-speed fan or hair dryer
wave tank
small pebbles
floating objects (Ping Pong balls, eyedroppers, corks, etc.)
coat hanger
Styrofoam ball (15 cm diameter)
Styrofoam ball (6 cm diameter)
dowel (1 m long and 0.5 cm diameter)
dowel (0.5 m long and 0.5 cm diameter)
2 weights (about 15 grams each)

Concept Area III

Activities:

Oily Spills
Forever Trash

water
vegetable oil or heavy olive oil (~ 500 ml)
salt (for salt water)
about 100 ml of liquid detergent
sand (~ 125 g)
sand containing organic matter
diatomaceous earth (~125 g)
cotton string (~ 1 meter)
feather
several small pieces of polystyrene foam (such as packing material)
small piece of paper (10 cm x 10 cm)
10-cm x 10-cm scraps of cloth (cotton, rayon, wool, polyester, nylon etc.)
small sheets of aluminum foil, waxed paper, plastic wrap
aluminum soda can pull tab
plastic bag – sandwich size
pieces of a plastic grocery bag (There are different kinds—some claim to degrade in light or landfills. Try to find an example of the different types.)
plastic six pack holder
plastic bottle cap
hard candy in a plastic wrapper
unwrapped hard candy
rubber balloon
polystyrene foam packing peanut
paper towels
several drinking straws (cut in half)
shoe box or small individual containers
dish pan or plastic tub
beaker

Resource Guide

This Resource Guide was compiled by the staff, consultants, and participants of Project Earth Science, and by the NSTA Press editors. It is not meant to be a complete representation of resources in oceanography, but will assist teachers in further exploration of this subject.

National Science Teachers Association
1840 Wilson Boulevard
Arlington, VA 22201-3000
(800) 277-5300
http://www.nsta.org/store

The National Science Teachers Association is the largest organization in the world committed to promoting excellence and innovation in science teaching and learning for all. NSTA's current membership of more than 53,000 includes science teachers, science supervisors, administrators, scientists, business and industry representatives, and others involved in science education. The NSTA Science Store has many resources on water and ocean studies.

National Sea Grant Office
NOAA/Sea Grant, R/ORI
1315 East-West Highway
SSMC-3, Eleventh Floor
Silver Spring, MD 20910
(301) 713-2448
http://www.nsgo.seagrant.org/

The National Sea Grant Program encourages the wise stewardship of our marine resources through research, education, outreach, and technology transfer. It serves as a partner and bridge between government, academia, industry, scientists, and private citizens to help Americans understand and sustainably use the Great Lakes and ocean waters for long-term economic growth. The National Sea Grant Program unites the National Oceanic and Atmospheric Administration (NOAA), 29 state Sea Grant programs, more than 200 universities, and millions of people. The Web site is a huge repository of educational publications, information on regional sea grant programs, and funding opportunities.

National Oceanic and Atmospheric Administration
14th Street and Constitution Avenue, NW
Room 6013
Washington, DC 20230
(202) 482-6090
http://www.noaa.gov/ocean.html

NOAA's Ocean page provides information about coral reefs, tides and currents, buoys, marine sanctuaries, estuaries, diving, oil and chemical spills, and links to various marine organizations both within and outside of NOAA. Also includes listings of NOAA publications and products. Especially worth visiting are the links to NOAA's National Ocean Service page and National Marine Sanctuaries page.

Bridge: Ocean Sciences Education Teacher Resource Center
Sea Grant Marine Advisory Services
Virginia Institute of Marine Science
Gloucester Point, VA 23062
http://www.vims.edu/bridge/

Bridge is a comprehensive Web site on marine science education resources. The goal of the site is to provide educators with content-correct and content-current marine information and data; to support researchers in outreach efforts; and to improve communications among educators and between the education and research communities.

The Environmental Protection Agency's Web site has an extensive link list of water-related sites appropriate for students and teachers.

Environmental Protection Agency
Student Center
Water Resource Page
1200 Pennsylvania Avenue, NW
Washington, DC 20460
http://www.epa.gov/students/
water.htm

The Center for Marine Conservation is the nation's leading non-profit organization dedicated solely to protecting the abundance and diversity of marine life. Publications and materials listed on their Web site are available for purchase through the Center's Marine Debris Information Office

Center for Marine Conservation
1725 DeSales Street, NW, Suite 600
Washington, DC 20036
(202) 429-5609
http://www.cmc-ocean.org/

The Woods Hole Oceanographic Institution is a private, independent, not-for-profit corporation dedicated to research and higher education at the frontiers of ocean science. Their Web site features an elaborate education section with educational kits, online resources, and information about their professional development programs.

Woods Hole Oceanographic Institution
Information Office
Co-op Building, MS #16
Woods Hole, MA 02543
(508) 289-2252
http://www.whoi.edu/home/

The South Slough National Estuarine Research Reserve is dedicated to the research, education, and stewardship of the 4,700 acre natural area that encompasses 600 acres of tidal marshes, mudflats and open water channels. Connecting to the ocean through the Coos estuary mouth, near Charleston, Oregon, South Slough provides an outstanding natural laboratory. The Web site presents the Estuaries Feature Series of articles on various marine subjects.

South Slough National Estuarine
Research Reserve
PO Box 5417
Interpretive Center - Seven Devils Road
Charleston, Oregon 97420
(541) 888-5558
http://www.southsloughestuary.com/

Project Oceanology is a marine science and environmental education organization operated by a non-profit association of schools, colleges, and other educational institutions in Connecticut, Massachusetts, Rhode Island, and New York. Project Oceanology provides boats, oceanographic equipment, a waterfront laboratory, instructional materials, and staff to enable more than 20,000 students and adults each year to learn about the ocean through first-hand, on-the-water experiences. The Web site also provides a list of publications and curriculum materials available for purchase.

Project Oceanology
Avery Point
1084 Shennecossett Road
Groton, CT 06340
(800) 364-8472
http://www.oceanology.org/

Pacific Science Center
200 Second Avenue North
Seattle, WA 98109
(206) 443-2870
http://www.pacsci.org/

Pacific Science Center is an independent, not-for-profit educational foundation dedicated to increasing the public's understanding and appreciation of science, mathematics, and technology through interactive exhibits and programs. Their Web site has an education section that includes professional development information and resource suggestions.

National Aquarium in Baltimore
Pier 3
501 E. Pratt Street
Baltimore, MD 21202
(410) 576-3800
http://www.aqua.org/

Committed to serving a diverse constituency, the National Aquarium in Baltimore seeks to stimulate interest in, develop knowledge about, and inspire stewardship of aquatic environments. The Web site provides information about museum programs, and online and purchasable teacher resource material.

Curriculum Research and Development Group
University of Hawai'i
1776 University Avenue
Honolulu, HI 96822
(800) 799-811
http://www.hawaii.edu/crdg/

The Curriculum Research & Development Group at the University of Hawai'i developed the *Fluid Earth/Living Ocean* curriculum for secondary school students of all abilities, to teach basic concepts of science through laboratory and field investigations into the marine environment. The content of the units is drawn from biology, physics, chemistry, meteorology, geology, cartography, oceanography, ecology, and marine engineering. The units may be taught as a complete marine science course, or they may be used as modules in other science courses.

Water Resources Division
U.S. Geological Survey
12201 Sunrise Valley Drive
Reston, VA 20192
(703) 648-4000
http://water.usgs.gov/education.html

The U.S. Geological Survey has the principal responsibility within the Federal Government to provide the hydrologic information and understanding needed by others to achieve the best use and management of the nation's water resources. Includes the Water Science for Schools site, which offers information on many aspects of water, along with pictures, data, maps, and an interactive center.

American Geophysical Union
Science for Everyone—Ocean Sciences
2000 Florida Avenue, NW
Washington, DC 20009-1277
(800) 966-2481
http://www.agu.org/sci_soc/
everyoneoc.html

The American Geophysical Union provides information for the public on what is known about Earth, its planetary neighbors, and the space around them and how the geophysical sciences are advancing. The Science for Everyone section provides articles, reports, and images especially selected for a general audience.

Society for Sedimentary Geology
1731 E. 71st Street
Tulsa, OK 74136-5108
(800) 865-9765
http://www.beloit.edu/~SEPM/

The Society for Sedimentary Geology's K–12 Earth Science Education Committee's Web site has free online activities on water and oceans.

This PBS site features educational resource materials that accompany the Secrets of the Ocean Realm video series and book. Online activities aimed at grades 5–7, covering topics such as oceanography, marine biology, ecology, physics, conservation, and scuba diving are also provided.

Secrets of the Ocean Realm
PBS/Corporation for Public Broadcasting
http://www.pbs.org/oceanrealm/

A companion to the *Savage Seas* video series, this Web site features sea facts, an ask-an-expert feature, links to other sites, and resource material available for purchase.

Savage Seas
PBS/Corporation for Public Broadcasting
http://www.pbs.org/wnet/savageseas/index.html

Great Explorations in Math and Science (GEMS) is a growing resource for the advancement of inquiry-based science and mathematics education. GEMS publishes more than 60 teacher's guides and handbooks, offers specialized workshop opportunities, and maintains a national network of teacher-training sites and centers. The Web site lists publications, products, and professional development opportunities.

GEMS
Lawrence Hall of Science
University of California, Berkeley
Berkeley, CA 94720-5200
(510) 642-7771
http://lhs.berkeley.edu/GEMS/

National Geographic's Web site for teachers and students has online activities, downloadable lesson plans, printable maps, and resource material available for purchase.

National Geographic Education
1145 17th Street, NW
Washington, DC 20036
(800) 368-2728
http://magma.nationalgeographic.com/education/index.cfm

The National Wildlife Federation's Education Web site features resources available for purchase or free download. The *NatureScope* series and *Animal Tracks* series features several titles dealing with water and oceans.

NatureScope/Animal Tracks
National Wildlife Federation
8925 Leesburg Pike
Vienna, VA 22184
(410) 516-6585
http://www.nwf.org/

Creative Teaching Associates is a producer of learning games and activities appropriate for both classroom and home use. Their science section includes several products about water.

Creative Teaching Associates
Fresno, CA 93747
(559) 291-6626
http://www.mastercta.com/

Constructing a Wave Tank

Constructing a wave tank is not overly difficult, although experience working with Plexiglas or building other projects will make the job easier. Give yourself plenty of time to work slowly and carefully. You might want to begin construction one to two weeks before scheduled classroom use to allow for construction and practice. Read all directions before beginning the assembly. You may also want to have extra materials on hand in case problems arise. (NOTE: Because the U.S. building industry still uses the English system of measurment, the instructions that follow will be in English units.)

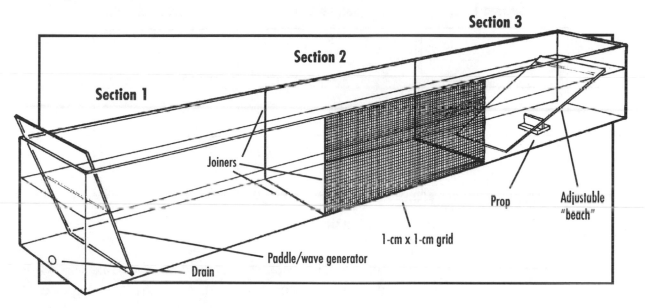

Figure 1

The assembled wave tank

Materials for Constructing 9-foot Wave Tank

◊ 9 pieces ¼" Plexiglas measuring 12" x 36" (S1A through S3B and B1 through B3)

◊ 2 pieces ¼" Plexiglas measuring 12" x 12¼" (E1 and E2)

◊ 1 piece ¼" Plexiglas measuring 11" x 15" (Paddle/Wave Generator)

◊ 1 piece ¼" Plexiglas measuring 11" x 24" (Beach)

◊ 1 piece ¼" Plexiglas measuring 4" x 4" (Prop1)

◊ 1 piece ¼" Plexiglas measuring 2" x 4" (Prop2)

◊ 6 transparent acrylic joiners: 4 @ 12" and 2 @ 11½"

◊ 4 acrylic hinges

◊ 1 tube acrylic cement

◊ power drill with bit size appropriate for drain plug
◊ 1 tube silicone sealant
◊ carpenter's square
◊ drain plug (about $\frac{1}{2}$" in diameter)
◊ overhead transparencies (3)

Procedure

1. The wavetank is assembled in three sections that may be taken apart for storage. See Figure 1. The Plexiglas should be purchased pre-cut to the specified sizes. Once you have assembled the necessary materials, arrange them as shown and label each piece of Plexiglas according to Figure 2.

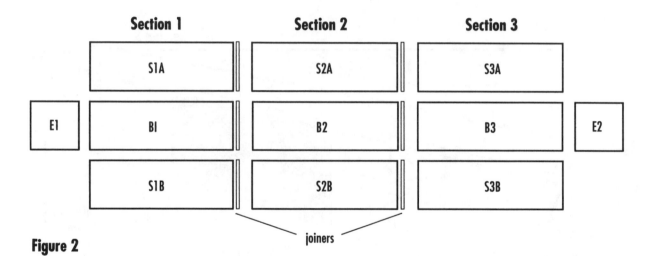

Figure 2

2. Into piece E1, carefully drill the drain hole. Take care when doing this so as not to crack the Plexiglas. The plug should fit snugly into the hole to prevent leakage. Position the hole as shown in Figure 3.

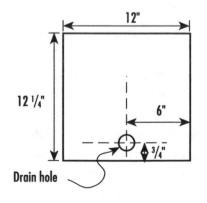

Figure 3

3. Carefully glue piece S1A to B1, using the carpenter's square to make certain that the pieces are square to one another. Put a line of the acrylic cement along the edge of piece S1A and glue the two pieces together as shown in Figure 4. Important: Make sure the two pieces are square to one another and the ends of the two pieces are even with one another.

Once S1A and B1 are firmly glued together, glue S2A to B2 and S3A to B3 as described above.

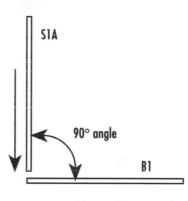

Figure 4

4. Carefully glue piece S1B to S1A/B1, using the carpenter's square to make certain that the pieces are square to one another. Put a bead of the acrylic cement along the edge of piece S1B and glue the two pieces together as shown in Figure 5. Important: Make sure the three pieces are square to one another and the ends of the three pieces are even with one another.

Once S1B and S1A/B1 are firmly glued together, glue S2B to S2A/B2 and S3B to S3A/B3 as described above.

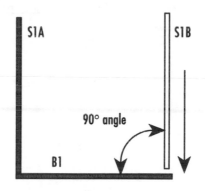

Figure 5

5. Carefully glue piece E1 to S1A/B1/S1B, using the carpenter's square to make certain that the pieces are square to one another. Put a bead of the acrylic cement along the edge of pieces S1A/B1/S1B and glue the four pieces together as shown in Figure 6.

Glue piece E2 to S3A/B3/S3B as described above.

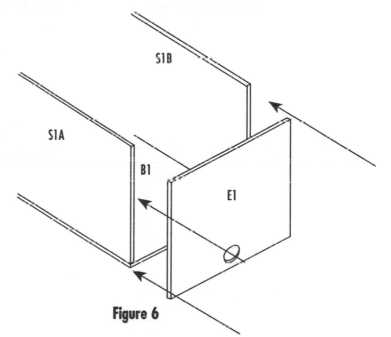

Figure 6

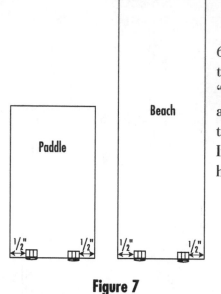

6. Carefully glue a pair of hinges to each of the two pieces of Plexiglas labeled "Beach" and "Paddle." Put a bead of the acrylic cement along one side of the hinge and glue the hinge to the Plexiglas as shown in Figure 7. Important: Be careful not to get glue into the hinging mechanism.

Figure 7

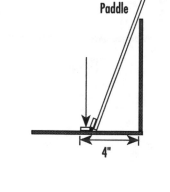

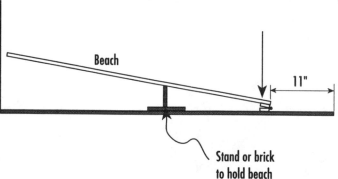

7. Via the hinges, carefully glue the "Paddle" to piece B1. Put a bead of the acrylic cement along the side of the hinge and glue the hinge to the Plexiglas as shown in Figure 8, making sure to center the "Paddle" within the tank, leaving approximately 1/4" between the "Paddle" and the sides of the tank. Important: Again, be careful not to get glue into the hinging mechanism. Repeat this procedure to attach the "Beach" to B3.

Figure 8

8. Using a bead of acrylic cement, glue the transparent joiners to the open end of Section 1 and to one end of Section 2 as shown in Figure 9. (See Figure 2 to identify the correct placement of the joiners.)

9. To make the prop for the adjustable beach, glue the pieces labeled "Prop 2" to "Prop 1" as shown in Figure 10. Use the carpenter's square to make sure the two pieces are square to one another.

10. After all pieces have had a chance to "set up" (about a day or so) put the wave tank on a counter, preferably with the drain positioned over a sink or close to a sink. This will make it easier to fill and drain the tank.

11. To assemble the three sections of the wave tank, simply fill the channels of the joiners on the end of one of the pieces of the tank with the silicone sealant . See Figure 11. Follow the directions for the application of the sealant on the package. You do not need to use an excessive amount.

12. Slide the wave tank pieces tightly together, fitting the end of the non-channeled piece of section 2 into the channel on the end of section 1. The sealant may squeeze out of the channel when the tank is properly connected. Repeat this procedure to join section 3 with sections 1 and 2. **Important: The sealant should be allowed to "set-up" overnight before filling the tank with water.**

13. Fill the tank with water to a depth of about 15 or 20 cm. Use a red wax marking pencil to note the locations of any leaks. Drain the tank and use the silicone sealant to seal any leaks. Allow the sealant to set up overnight before refilling the tank and using it with students.

14. Create 3 transparent grids by transferring a 1-cm x 1-cm grid from a lab manual to overhead transparencies. Place the grids on the side of the tank, securing them to Section 2 with removable transparent tape as in Figure 1.

15. After use, the tank may be disassembled for storage. Drain the tank and pull apart the three sections. Remove any excess sealant from the channel and from the ends of the tank.

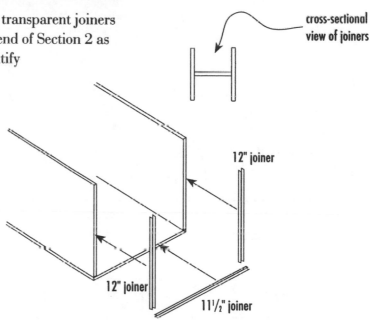

cross-sectional view of joiners

12" joiner

12" joiner

11½" joiner

Figure 9

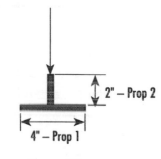

2" – Prop 2

4" – Prop 1

Figure 10

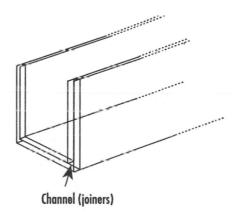

Channel (joiners)

Figure 11

Activity #	Subject Matter	Scientific Inquiry	Unifying Concepts and Processes
Activity 1	nature of water molecules		form and function
Activity 2	water molecules, polarity, and hydrogen bonding		evidence and models
Activity 3	water density solid and liquid states	tools and techniques to gather data	form and function
Activity 4	specific heat capacity of sand and water	tools and techniques to gather and analyze data	constancy, change, and measurement
Activity 5	solubility		form and function
Activity 6	measuring density of frresh and salt water		constancy, change, and measurement
Activity 7	ocean layers		evidence, models, and explanation
Activity 8	different densities of ocean layers		evidence, models, and explanation
Activity 9	nature and importance of estuaries		evidence, models, and explanation
Activity 10	how land and wind affect currents		evidence, models, and explanation
Demo 11	wave motion and energy		constancy, change, and measurement
Activity 12	wind's influence on waves		evidence, models, and explanation
Demo 13	modeling wave formation		evidence, models, and explanation
Activity 14	tide patterns		systems, order, and organization
Activity 15	how tides are formed		evidence, models, and explanation
Demo 16	model the moon's affect on tides		evidence, models, and explanation
Activity 17	oil spills and oceans		form and function
Activity 18	decomposition of ocean pollution		evidence, models, and explanation

Technology	Personal/Social Perspectives	Earth and Space Science Content	Activity Name
		structure of water molecules	**A Pile of Water**
		structure of water molecules	**A Sticky Molecule**
		structure and behavior of water	**Over and Under—Why Water's Wierd**
		properties of water vs. sand	**How Water Holds Heat**
		structure and properties of water	**Water—The Universal Solvent**
build a hydrometer		properties of water	**Won't You BB My Hydrometer**
		structure of Earth's systems	**Ocean Layers**
	science in society	structure of Earth's systems	**The Myth of Davy Jones's Locker**
		structure of Earth's systems	**Estuaries—Where Rivers Meet the Sea**
		structure of Earth's systems, Earth's history	**Current Events in the Ocean**
		structure of waves	**Body Waves**
build a wave tank		structure of Earth's systems	**Waves and Wind in a Box**
measure waves with a wave tank		structure of waves	**Tanks A Lot**
	tsunamis as a natural hazard	structure of Earth's systems	**Plotting Tidal Curves**
build a planet/tide model		Earth in the solar system	**Tides Mobile**
build an inertia/tide model		Earth in the solar system	**The Bulge on the Other Side of Earth**
technology behind cleaning oil spills	risks and benefits	structure of Earth's system	**Oily Spills**
	risks and benefits	structure of various materials vs. water	**Forever Trash**

*sci*LINKS

Project Earth Science: Physical Oceanography brings you *sci*LINKS, a creative project from NSTA that blends the best of the two main educational "drivers"—textbooks and telecommunications—into a dynamic new educational tool for all children, their parents, and their teachers. This *sci*LINKS effort links specific textbook and supplemental resource locations with instructionally rich Internet resources. As you and your students use *sci*LINKS, you'll find rich new pathways for learners, new opportunities for professional growth among teachers, and new modes of engagement for parents.

In this *sci*LINKed text, you will find an icon near several of the concepts you are studying. Under it, you will find the *sci*LINKS URL (http://www.scilinks.org/) and a code. Go to the *sci*LINKS Web site, log in, type the code from your text, and you will receive a list of URLs that are selected by science educators. Sites are chosen for accurate and age-appropriate content and good pedagogy. The underlying database changes constantly, eliminating dead or revised sites or simply replacing them with better selections. The ink may dry on the page, but the science it describes will always be fresh.

The selection process involves four review stages:
1. First, a cadre of undergraduate science education majors searches the World Wide Web for interesting science resources. The undergraduates submit about 500 sites a week for consideration.
2. Next, packets of these Web pages are organized and sent to teacher-Webwatchers with expertise in given fields and grade levels. The teacher-Webwatchers can also submit Web pages that they have found on their own. The teachers pick the jewels from this selection and correlate them to the *National Science Education Standards*. These pages are submitted to the *sci*LINKS database.
3. Then scientists review these correlated sites for accuracy.
4. Finally, NSTA staff approves the Web pages and edits the information provided for accuracy and consistent style.

Who pays for *sci*LINKS? *sci*LINKS is a free service for textbook and supplemental resource users, but obviously someone must pay for it. Participating publishers pay a fee to NSTA for each book that contains *sci*LINKS. The program is also supported by a grant from the National Aeronautics and Space Administration (NASA).